Doubly Classified Model with R

Teck Kiang Tan

Doubly Classified Model with R

 Springer

Teck Kiang Tan
Centre for Skills, Performance
   and Productivity
Institute for Adult Learning
Singapore
Singapore

ISBN 978-981-13-4981-2   ISBN 978-981-10-6995-6   (eBook)
https://doi.org/10.1007/978-981-10-6995-6

Printed on acid-free paper

This Springer imprint is published by Springer Nature
The registered company is Springer Nature Singapore Pte Ltd.
The registered company address is: 152 Beach Road, #21-01/04 Gateway East, Singapore 189721, Singapore

Dedicated to my late parents, Tan Eng Huat and Long Siew Long, and sister Tan Lee Kiang who supports me in all ways throughout in my life.

Foreword

Richard Branson once said, "There is no greater thing you can do with your life and work than to follow your passions in a way that serves the world and you." As I supervised Teck as a graduate student and later observed his contributions to the world of research, I can truly see how this book is his heartfelt effort to help aspiring statisticians leverage on the affordances of statistics.

He has done this in two profound ways. First, he noticed that despite the usefulness of doubly classified models for categorical data analysis, it was under-discussed in textbooks. This led him to elaborate on a wider range of relevant models; going beyond the commonly discussed models like independence and quasi symmetry models. Second, as an active researcher, he realized that beyond understanding the range of models available, the next step for any interested statistician would be to generate models for practical applications. Unfortunately, this would involve the use of statistical software like SAS or SPSS that is both expensive and not easily accessible. Cognizant that not everyone has access to research funds or software, Teck has taken pains to develop new syntax for generating model on Package R, a free statistical software that is gaining widespread popularity.

I am glad that Teck has written this book because it is not just about explaining the intricacies of one double classified model over another but it is to help interested students get an entry pass into this world and its possibilities. For example, Teck understands that not everyone is comfortable with mathematical symbols and has developed a "symbolic table" to help represent notations that are more complex. This symbolic table is a graphical tool to help readers understand the various models discussed.

Although this book gives many tips and guidelines on how to develop double classified models, I can see how this book may also be useful for readers who are interested in learning about Package R and use the guided steps as a lesson of R's functionalities. Whatever your purpose, I am sure that you would appreciate the examples included to illustrate concepts and its applications.

Trivina Kang
Office of Teacher Education (OTE),
Policy & Leadership Studies (PLS),
National Institute of Education
Nanyang Technological University
Singapore

Preface

The motivation in writing this book comes after the presentation of a poster at the R conference 2015 held at Denmark, Alborg, titled "Extending the Quasi-Symmetry Model: Quasi-Symmetry with n Degree." I found that modeling doubly classified table is quite unknown to most conference attendees. Although there has been a dramatic growth in the development and application of doubly classified models, there is not a single book that is written in R on this subject. The applications on doubly classified modeling are also restricted to illustrate it using software like SPSS and SAS, appeared most often in journal articles (e.g., Lawal and colleagues). Doubly classified models are not commonly appeared in statistic textbooks. Even there are books written on this subject, it is often restricted to a section or at most a chapter. This is probably the right time to put these models in an organized way into a book for sharing.

This book focuses on doubly classified models. The main aim of the book is to describe doubly classified models in a way readers can easily understand. Although mathematical representations are unavoidable for purpose of clarity, they are always accompanied with explanation in plain language. As doubly classified is probably not a familiar topic to a lot of people, including statisticians and data analysts, for those who have not heard of it, this book serves you well. You will find it a good starting point to have a basic understanding about doubly classified models. A lot of examples accompanied the explanation are given in the text to illustrate the characteristics of the various doubly classified models. It is applied oriented however not losing its function as a basic textbook and reference. The formulas for the various doubly classified models are given, together with examples illustrating the concepts and usages. For those who are not familiar with doubly classified modeling will find this book easier to understand. A new presentation form, referred to as symbolic table, is used throughout the text as a summarized table to explain the main characteristics of doubly classified model. Researchers, data analysts, and undergraduate and postgraduate students are suited audiences and most suitable for applied statisticians and researchers aim to use doubly classified model as an analytical tool for their studies.

Singapore, Singapore   Teck Kiang Tan
September 2017

Acknowledgements

I could not have written this book without the help and support of many people. I would like to express my appreciation for people who one way or the other who provided comment, help, and assistance in preparing the book. First and foremost, I like to thank Trivial Kang for the foreword; Johnny Sung, Sim Soo Kheng, Emily Low, and Regina Tan for their administrative support to make this possible. I also like to thank Goh Kiah Mok, John Wang, and Melvin Chan for their comments to improve the book.

Contents

Symbols

n_{ij}	Observed frequency for cell(i, j)
m_{ij}	Expected frequency for cell(i, j)
$\hat{m}_{ij}$	Fitted frequency for cell(i, j)
l_{ij}	Log of estimated frequency, $\ln(\hat{m}_{ij})$
π_{ij}	Probability for cell(i, j)
Δ_{upper}	Expected group cells of upper diagonal
Δ_{lower}	Expected group cells of lower diagonal
Ω	Odds $\frac{\pi}{1-\pi}$
δ_k	$\delta_k = \pi_{ij}/\pi_{ji}$ where $k = j - i$
τ	Log of odds, $\ln(\delta)$
θ_{ij}	Odds ratio, $\hat{\theta}_{ij} = \frac{\hat{m}_{ij}\hat{m}_{i+1,j+1}}{\hat{m}_{i,j+1}\hat{m}_{i+1,j}}$
Φ_{ij}	Log of odds ratio $\ln(\theta_{ij})$
$\Omega_{(ij,st)}$	$\frac{\Theta_{(ij,st)}}{\Theta_{(st,ij)}}$
$\Theta_{(ij,st)}$	$\frac{\pi_{is}\pi_{jt}}{\pi_{js}\pi_{it}}$

Chapter 1
Introduction to R

1.1 What Is R?

R is a programming language. Also, a language with "standard" syntax specified in a simple format to produce standard statistical output. This is done by using the functions specified in the various R packages. R provides a wide variety of statistical and graphical functions to carry out statistical analyses and produce elegant graphical outputs. The statistical models available in R include linear and nonlinear modeling, classical statistical tests, time series analysis, classification, clustering analysis, text mining, and propensity score analysis, just to mention a few of them. Since its inception in the middle of the 1990s, it has grown exponentially and now became a popular software.

1.2 Libraries in R

R consists of two parts: the base system provides the core R language and the add-on packages. While the core provides the fundamental libraries, the add-on packages are written by R users, which consist of a list of functions for a package to perform specific tasks, usually cover a specific content domain. For instance, package exact2 $\times$ 2 provides functions to carry out exact conditional tests for 2 $\times$ a tables that summarize count data, and packages ggplot2 provides elegant data visualization based on the principal of grammar of graphics.

R is a freeware. It can be downloaded and installed from Comprehensive R Archive Network (CRAN). Upon installation of R package, the base system is loaded, i.e., a core set of functions are in existence after the installation. This base package offers the basic functions to perform statistical modeling such as linear model [lm() function] and generalized linear model [glm() function]. Add-on packages are separately installed and have to load them during processing in order

© Springer Nature Singapore Pte Ltd. 2017
T.K. Tan, *Doubly Classified Model with R*,
https://doi.org/10.1007/978-981-10-6995-6_1

1

to use them. These are packages written by R users. The scope and content area are wide and coverage is extensive. The add-on packages range from specialized statistical models, data mining, graphical devices, import/export capabilities, parallel processing to reporting tools. As at March 2, 2017, there are 10,187 add-on packages available at the Comprehensive R Archive Network (CRAN).

The following Table 1.1 lists a few examples of the add-ons R packages. The lattice package is a graphic library that produces lattice graphic. The MASS package contains data sets and functions written by Venables and Ripley for the book "Modern Applied Statistics" (MASS). The foreign package contains functions for reading and writing data stored in different formats such as Epi Info, Minitab, S, SAS, SPSS, Stata, Systat, and Weka to read in R data. It also includes functions for reading and writing some dBase files. The xlsx package provides R functions to read, write, and format Excel 2007 and Excel 97/2000/XP/2003 file formats.

Each add-on package is not stand-alone. The functions in the add-on packages can be interrelated to other add-on packages. This interconnectedness of functions is essential as using R to carry out analysis normally required using more than one library and output from one function of one package can be read in into another function of another package. The list of available CRAN packages sorted by name is available at the CRAN Web page. Refer to the CRAN Web page below for the add-on packages.

http://cran.r-project.org/web/packages/available_packages_by_name.html

Table 1.1 Examples of R packages

Package	Description
car	Companion to applied regression
catspec	Special models for categorical variables
descr	Descriptive statistics
epitools	Epidemiology tools
exact2 × 2	Exact conditional tests for 2 by 2 tables of count data
fmsb	Functions for medical statistics book with some demographic data
foreign	Read data files from statistical packages, e.g., SAS, SPSS, Stata
ggplot2	Elegant data visualization using the *Grammar of Graphics*
gmodels	Various R programming tools for model fitting
gnm	Generalized nonlinear model
gridExtra	Miscellaneous functions for "Grid" graphics
hypergea	Hypergeometric tests
lattice	Lattice graphics for panel plots or trellis graphs
latticeExtra	Extra graphical utilities based on lattice
MASS	Package associated with Venables and Ripley's book entitled Modern Applied Statistics using S-PLUS
psych	Procedures for psychological, psychometric, and personality research
reshape	Flexibly reshape data
reshapes	Flexibly reshape data—A reboot of the reshape package
vioplot	Violin plot
xlsx	Read, write, format Excel 2007, and Excel 97/2000/XP/2003 files

1.3 Installing R

R is available as free software under the terms of the Free Software Foundation's GNU General Public License in source code form. It compiles and runs on UNIX platform, Windows, and MacOS and can be installed from the Web site Comprehensive R Archive Network (CRAN). The address is as follows:

http://cran.r-project.org/

Alternatively, you can search from Google by typing "CRAN" to locate the site. The Web page is shown below. Once you are there, follow the instructions indicated in the Web page and you will be able to download R. Store the downloaded installation file in a directory and run it to install R.

CRAN
Mirrors
What's new?
Task Views
Search

About R

The Comprehensive R Archive Network

Download and Install R

Precompiled binary distributions of the base system and contributed packages, **Windows and Mac** users most likely want one of these versions of R:

- Download R for Linux
- Download R for (Mac) OS X
- Download R for Windows

1.4 Running R

After installation, you will see an icon indicating the version of R package as shown below. As of April 21, 2017, the version of R is Version 3.4.0. R version upgrade is usually fast. So, always look for upgrading to a newer version to take advantages of latest functions available. To run R, simply click on the R icon.

After clicking on the icon, you will get the following screen.

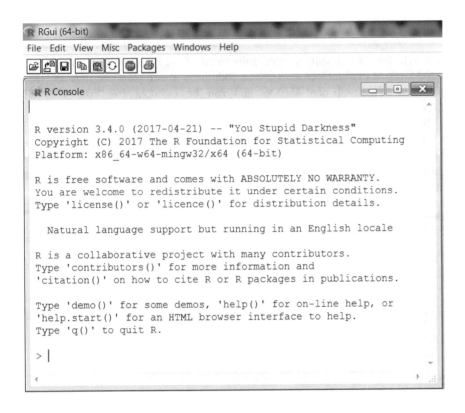

The command prompt (>) in red is where we can execute R command. For instance, by typing 2 + 3 after the > , the screen will return 5.

1.5   Installing Add-on Packages and Libraries

As mentioned earlier, after installation of R, the base package is automatically loaded when the R package is invoked. However, the base package of R does not contain the add-on packages/libraries written by R users. In order to use these packages, they have to be downloaded and installed. There are two ways to install add-on packages. One is to enter the command install.packages() at the command prompt > to execute the process of downloading the libraries. The syntax below installs two add-on packages: foreign and xlsx.

```
install.packages("foreign")
install.packages("xlsx")
```

Another way to install the above two packages is to use the menu in Package R. Click on <Packages> and <Install package(s)> , the <HTTPS CRAN mirror> screen will pop-up showing a list of countries in alphabetic order. Select one of the countries to indicate where the package you want to download from and click <OK>. Another screen <Packages> will pop-up with a list of all the available packages in alphabetical order. Browse through the list, select the package you want to install, and click <OK> to execute the installation.

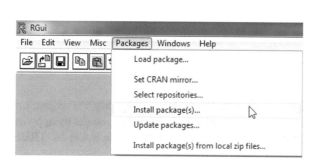

1.6   Basic Data Types

There are several basic R data types that are of frequent occurrence in routine R calculations. The following lists five of them and the subsections elaborate their usages.

1. Numeric
2. Integer
3. Logical
4. Character
5. Date

1.6.1   Numeric

Decimal values are referred to as numeric in R. It is the default computational data type. If we assign a decimal value, say 10.5 to a variable x, x is automatically considered as a numeric type. If we assign an integer to a variable x, say 10, it is still referred to as a numeric value. The class() function returns the data type of a vector. The is.integer() command tests whether a given vector is a numeric. If it is numeric, it returns TRUE, if not, then returns FALSE.

```
> x1 <- 10.5
> class(x1)
[1] "numeric"
> x2 <- 10
> class(x2)
[1] "numeric"
> # Is x2 an integer?
> as.integer(x2)
[1] 10
> class(x2)
[1] "numeric"
> is.integer(x1)
[1] FALSE
> is.integer(x2)
[1] FALSE
```

1.6.2   Integer

In order to create an integer variable in R, we have to specify the as.integer() function. The following example assures vector y1 is indeed an integer by applying the as.integer() function. The is.integer() function returns the value of TRUE, indicating y1 is an integer.

```
> y1 <- as.integer(5)
> class(y1)
[1] "integer"
> is.integer(y1)  # is y1 an integer?
[1] TRUE
```

We can coerce a numeric value into an integer using the as.integer() function. Similarly, we could also parse a string for decimal values the same way.

```
> as.integer(5.14)    # coerce a numeric value
[1] 5
> as.integer("5.67")  # coerce a decimal string
[1] 5
```

However, an error will occur if we try to parse a non-decimal string.

```
> as.integer("Mary")  # coerce an non-decimal string
[1] NA
Warning message:
NAs introduced by coercion
```

Sometimes, it is useful to perform arithmetic on logical values. TRUE returns a value 1, while FALSE returns a value 0.

```
> as.integer(TRUE)    # the numeric value of TRUE
[1] 1
> as.integer(FALSE)   # the numeric value of FALSE
[1] 0
```

1.6.3   Logical

A logical value is often created via comparison between variables. The following example shows that after assigning 1 to a, and 2 to b, the condition of a > b, stored in h, returns as FALSE. The class of h is a logical vector.

```
> a <- 1
> b <- 2
> h <- a > b        # Is a larger than b?
> h                 # Print the logical value
[1] FALSE
> class(h)
[1] "logical"
```

Standard logical operations "&" (and), "|" (or), and "!" (Negation) could be used for logical data type. The following example shows that u is assigned with TRUE and v assigned as FALSE, the result of u & v returns as FALSE while u|v returns as TRUE. The negation of u (!u) and v (!v) results in FALSE and TRUE, respectively.

```
> u <- TRUE
> v <- FALSE
> u & v             # u AND v
[1] FALSE
> u | v             # u OR v
[1] TRUE
> !u                # Negation of u
[1] FALSE
> !v                # Negation of V
[1] TRUE
```

The following Table 1.2 summarizes the logical operations used in base R.

1.6.4   Character

A character object in R stores the string values. A character object can be easily created using " " as shown below.

```
> a <- "Doubly Classified Model"
> a
[1] "Doubly Classified Model"
> class(a)
[1] "character"
```

Table 1.2 Logical operators

Operator	Description
<	Less than
<=	Less than or equal to
>	Greater than
>=	Greater than or equal to
==	Exactly equal to
!=	Not equal to
!x	Not x
x| y	x OR y
x & y	x AND y
is.TRUE(x)	Test if x is TRUE

Objects can be converted into character values with the as.character() function. The following syntax converts the value 5.6 into a character "5.6."

```
> x <- as.character(5.6)
> x
[1] "5.6"
> class(x)
[1] "character"
```

Two character values can be concatenated with the paste() function.

```
> fname <- "Kang"
> lname <- "Trivina"
> paste(lname, fname)
[1] "Trivina Kang"
```

It is often more convenient to create a readable string with the sprintf () function.

```
> sprintf("%s has %d dollars", "Ruby", 100)
[1] "Ruby has 100 dollars"
```

The sprintf() function prints Chinese characters as well.

```
> sprintf("一分耕耘一分收穫")
[1] "一分耕耘一分收穫"
```

The substr() function extracts a substring from a string. The following is an example showing how to extract the substring between the third and twelfth positions of a string.

```
> substr("Mary has a little lamb.", start=3, stop=12)
[1] "ry has a l"
```

To replace the first occurrence of the word "little" by another word "big" in the string, we apply the sub() function.

```
> sub("little", "big", "Mary has a little lamb.")
[1] "Mary has a big lamb."
```

1.7 Data Structure

R has a wide variety of data structures. The basic data structures include scalars, vectors (numerical, character, logical), matrices, data frames, lists, and factors. They are briefly discussed in the following subsections.

1.7.1 Vector

The most basic data structure is a vector. A vector is a sequence of data elements of the same basic type. Members in a vector are called components. The simplest structure of vector is probably the numeric vector, which is a single entity consisting of an ordered collection of numbers.

1.7.1.1   Vector Assignment

R stores data with a name given to it. To set up a numeric vector named x consisting of five components, the command is as follows:

```
x <- c(1,3,2,7,5)
```

Notice that the assignment operator (<−), which consists of the two characters "<" (less than) and "−" (minus) occurring strictly side-by-side and it points to the object receiving the value of the expression. In most contexts, the "=" operator can be used as an alternative. The following syntax is equivalent to the previous syntax to state the vector x.

```
x = c(1,3,2,7,5)
```

Vector assignments can also be made in the other direction by changing the arrow the other way in the assignment operator as shown below. However, this is not encouraged as it can be confusing and generally viewed as "bad" documentation.

```
c(1,3,2,7,5) -> x
```

The following is a vector y consisting of only logical values.

```
y <- c(TRUE, FALSE, TRUE, FALSE, FALSE)
```

A vector can contain character strings as well.

```
z <- c("aa", "bb", "cc", "dd", "ee")
```

To find out the number of members in a vector, use the length() function. There are five elements in vector z.

```
> length(z)
[1] 5
```

1.7.1.2   Combining Vectors

Vectors can be combined via the concatenation function c(). For example, the following two vectors n and c are combined into a new vector called nc containing elements from both vectors.

```
n <- c(2, 3, 5)
c <- c("aa", "bb", "cc", "dd", "ee")
nc <- c(n,c)
```

```
> nc
[1] "2"  "3"  "5"  "aa" "bb" "cc" "dd" "ee"
> class(nc)
[1] "character"
```

The above example shows that when numeric and character vector are combined, the numeric values are being coerced into character strings.

1.7.1.3 Vector Arithmetic

Arithmetic operations of vectors are performed member-by-member, i.e., memberwise according to the position of the elements in the vector. The following example shows that when we multiply vector a by 10, we would get a vector with each of its members multiplied by 10.

```
a <- c(2, 4, 6, 8)
b <- c(1, 2, 4, 8)
```

```
> c <- 10 * a
> c
[1] 20 40 60 80
```

If we add vector a and vector b together, the sum would be a vector whose members are the sum of the corresponding members from vector a and vector b.

```
> d <- a + b
> d
[1]  3  6 10 16
```

Similarly for subtraction, multiplication, and division, we get new vectors via memberwise operations.

```
> e <- a - b
> e
[1] 1 2 2 0
> f <- a * b
> f
[1]  2  8 24 64
> g <- a / b
> g
[1] 2.0 2.0 1.5 1.0
```

1.7.1.4 Recycling Rule

If two vectors are of unequal length, the shorter one will be recycled in order to match the longer vector. The following vectors u and v have different lengths, and their sum is computed by recycling values of the shorter vector u.

```
> u <- c(10, 20, 30)
> v <- c(1, 2, 3, 4, 5, 6, 7, 8, 9)
> u + v
[1] 11 22 33 14 25 36 17 28 39
```

1.7.1.5   Vector Index

After creating a vector, we may want to use a section or few sections of the elements. This can be carried out by declaring an index inside the square bracket "[]" operator to extract them. The following six examples show how to retrieve values from a vector using the [] operator. There are 14 element values stored in vector data1. To retrieve the fourth element of data1, specify data1[4]. Similarly, to retrieve the third to the sixth elements, instead of using the command data1[c (3,4,5,6)]. A shorter command data1[3:6] will also produce the same result. If we want to select the third, sixth, and seventh elements, specify data1[c(3,6,7)]. We could also use the minus sign "−" to exclude elements we want to eliminate. Specifying the command data1[−c(1:3)] excludes the first to the third elements. To exclude more than a list, we insert "," and specify the second list of elements after the first list; data[−c(1:3,11:14)] excludes two lists of elements. Comparative operators could be used to extract data; data1[data1 > 6] specifies elements with value greater than 6 are excluded.

```
data1 <- c(2,2,2,3,3,3,4,5,6,7,8,8,8,8)
data1[4] # The forth element in data1
data1[3:6] # Include thirdto sixth elements
data1[c(3,6,7)] # Include third, sixth and seventh elements
data1[-c(1:3)] # Exclude first to third elements
data1[-c(1:3,11:14)] # Exclude first to third and eleventh to fourteen
elelments
data1[data1>6] # Elements greater than 6
```

```
> data1[4]                # The forth element in data1
[1] 3
> data1[3:6]              # Include thirdto sixth elements
[1] 2 3 3 3
> data1[c(3,6,7)]        # Include third, sixth and seventh elements
[1] 2 3 4
> data1[-c(1:3)]          # Exclude first to third elements
 [1] 3 3 3 4 5 6 7 8 8 8 8
> data1[-c(1:3,11:14)]   # Exclude first to third and eleventh to fourteen elelments
[1] 3 3 3 4 5 6 7
> data1[data1>6]         # Elements greater than 6
[1] 7 8 8 8 8
```

1.7.1.6   Duplicate Indexes

The index vector allows duplicate values as well. Hence, the following example retrieves the seventh member twice.

```
> Duplicate <- data1[c(3, 7, 7)]
> Duplicate
[1] 2 4 4
```

1.7.1.7 Out-of-Order Indexes

The index vector can be out-of-order. The following shows a vector order is created that is not in numeric order. The seventh element is first extracted followed by the third and tenth element.

```
> Order <- data1[c(7, 3, 10)]
> Order
[1] 4 2 7
```

1.7.2 Matrix

A matrix is a collection of data elements arranged in a two-dimensional rectangular layout. The following is an example of a matrix with two rows and three columns.

$$A = \begin{bmatrix} 1 & 2 & 3 \\ 2 & 5 & 7 \end{bmatrix}$$

The following shows the creation of a matrix A using the matrix() function. The byrow = TRUE specifies the matrix is read in by row with two rows (nrow = 2) and three columns (ncol = 3). It is noted that the data elements of a matrix must be of the same basic type.

```
A <- matrix(
c(1, 2, 3, 2, 5, 7), # the data elements
nrow=2, # number of rows
ncol=3, # number of columns
byrow = TRUE) # fill matrix by rows
```

```
> A
     [,1] [,2] [,3]
[1,]   1    2    3
[2,]   2    5    7
```

An element at the mth row, nth column of A can be extracted by specifying the row number m and column number n, A[m, n]. The following extracts the second row element and third column element from matrix A

```
> A[2,3]
[1] 7
```

The entire mth row of matrix A can be extracted as A[m,].

```
> A[2,]
[1] 2 5 7
```

Similarly, the entire nth column of matrix A can be extracted with the syntax A[,n].

```
> A[,3]
[1] 3 7
```

We can also extract more than one rows or columns at a time. The following specifies the entire row of matrix A, and only the first and third column of matrix A is extracted.

```
> A[,c(1,3)]
     [,1] [,2]
[1,]    1    3
[2,]    2    7
```

1.7.3   Data Frame

If data are organized by column and row where row contains observations and column contains variables, data frame is the most suitable data structure. It could be viewed as a list of vectors of equal length that grouped together to form a rect-angular table structure. Most statistical software like SAS, SPSS, and STATA organize their data in data frame format. The direct way to create a data frame is to use the data.frame() function. The following illustrates a data frame named as dataframe1 which is created by reading in three vectors n, s, and l, each with five observations. By specifying data.frame(n,s,l), the numeric vector n, character vector s, and logical vector l are grouped into a data frame called dataframe1.

```
n <- c(2, 3, 5, 6, 7)
s <- c("aa", "bb", "cc", "dd", "ee")
l <- c(TRUE, FALSE, TRUE, TRUE, TRUE)
dataframe1 <- data.frame(n, s, l)
dataframe1
```

```
> dataframe1
  n  s     l
1 2 aa  TRUE
2 3 bb FALSE
3 5 cc  TRUE
4 6 dd  TRUE
5 7 ee  TRUE
```

To view what is stored in the cell of the first row and second column of data-frame1, use the following command:

```
> dataframe1[1,2]
[1] aa
Levels: aa bb cc dd ee
```

To identify how many row and columns are there in the data frame, use the nrow() and ncol() commands.

```
> nrow(dataframe1)
[1] 5
> ncol(dataframe1)
[1] 3
```

Subsetting data into smaller set is common. The following shows how data frame could be subset using the subset() command. By specifying the logical operation in subset, the following subset() command restricts dataframe1 to variable n with values greater than five and stores it to a new data frame called Subset1.

```
# Subset Row
Subset1 <- subset(dataframe1,n>=5)
Subset1
```

```
> Subset1
  n s  l
3 5 cc TRUE
4 6 dd TRUE
5 7 ee TRUE
```

The following subset() and select() commands subset dataframe1 to include only variable s and l and store it to a new data frame Subset2.

```
# Subset Column
Subset2 <- subset(dataframe1, select=c(s,l))
Subset2
```

```
> Subset2
   s    l
1 aa  TRUE
2 bb FALSE
3 cc  TRUE
4 dd  TRUE
5 ee  TRUE
```

Subsetting both row and column can be performed together. The syntax below subsets dataframe1 by row and column and stores to a new data frame Subset3.

```
# Subset Row and Column
Subset3 <- subset(dataframe1, n>=5, select=c(s,l))
Subset3
```

```
> Subset3
   s    l
3 cc TRUE
4 dd TRUE
5 ee TRUE
```

Another way of subsetting data frame is to use the bracket notation [] which makes use of index for subsetting. These indices refer to the number specified in the []. The first index is referring to the rows and the second for the columns. The first index is the number specified before the "," and the second index after it. Subset4 specifies only the first and second columns of dataframe1 are selected, whereas Subset5 specifies only the first and second rows of dataframe1 are selected.

```
Subset4 <- dataframe1[,1:2] # Column Subset
Subset4
Subset5 <- dataframe1[1:2,] # Row Subset
Subset5
```

```
> Subset4
  n s
1 2 aa
2 3 bb           > Subset5
3 5 cc             n s    l
4 6 dd           1 2 aa  TRUE
5 7 ee           2 3 bb FALSE
```

The indices in [] could be used together with the "==", %in%, and c() com-mands. Using the "==" command, Subset6 is created by restricting the data frame which contains those observations that contain only TRUE values of variable l. Similarly, %in% command can be used together with the concatenate function c() to subset values with 2, 5, and 7 of variable n into the new data frame Subset7. Subset8 data frame further restricts datafram1 to contain only variables n and s.

```
Subset6 <- dataframe1[dataframe1$l == TRUE,]
Subset6
Subset7 <- dataframe1[dataframe1$n %in% c(2,5,7),]
Subset7
Subset8 <- dataframe1[dataframe1$n %in% c(2,5,7),1:2]
Subset8
```

```
> Subset6
  n s    l         > Subset7            > Subset8
1 2 aa TRUE           n s    l            n s
3 5 cc TRUE         1 2 aa TRUE         1 2 aa
4 6 dd TRUE         3 5 cc TRUE         3 5 cc
5 7 ee TRUE         5 7 ee TRUE         5 7 ee
```

1.7.4   List

While matrix stores data in two-dimensional spaces of the same data type and data frame allows different data types to store in different columns in a two-dimensional table, a list is more flexible than matrix and data frame that can vary in length and also store data with different data types. A list can possibly consist of a numeric vector, a logical value, a matrix, a complex vector, a character array, and a function. For example, the following list1 is a list containing of three vectors n, s, l, and a numeric value 3. The output of a generalized linear model is stored in a lm or glm class.

```
n <- c(2, 3, 5)
s <- c("aa", "bb", "cc", "dd", "ee")
l = c(TRUE, FALSE, TRUE, FALSE, FALSE, TRUE)
list1 <- list(n, s, l, 3)
list1
```

```
> list1
[[1]]
[1] 2 3 5

[[2]]
[1] "aa" "bb" "cc" "dd" "ee"

[[3]]
[1]  TRUE FALSE  TRUE FALSE FALSE  TRUE

[[4]]
[1] 3
```

The components of a list can be retrieved using []. By specifying list1[2], it retrieves the second component from list1. To retrieve the first and second components, change the command to list1[c(1,3)].

```
> list1[2]
[[1]]
[1] "aa" "bb" "cc" "dd" "ee"

> list1[c(1,3)]
[[1]]
[1] 2 3 5

[[2]]
[1]  TRUE FALSE  TRUE FALSE FALSE  TRUE
```

1.7.5 Factor

A factor is ideal for specifying categorical data, in particularly for the purpose of regression modeling. Conceptually, factors are variables which take on a limited number of different values; such variables are often referred to as dummy variables in regression analysis. One of the most important uses of factor in statistical modeling is in regression analysis to specify a dummy variable. As the characteristics of a categorical variable are different from that of a continuous variable, stipulating a categorical variable as a factor into a linear model is a straightforward specification for regression analysis.

The first two lines of the syntax below create a character vector Race. The as.factor() function coerces its argument to a factor, named as RaceFactor. Race is stored in character as the output printed with "" for each of the elements while RaceFactor is factor, the output for each element is printed without "".

```
Race <- rep(c("Chinese", "Malay", "Indian", "Others"), each=6)
Race
class(Race)
RaceFactor <- as.factor(Race)
RaceFactor
class(RaceFactor)
```

```
> Race
 [1] "Chinese" "Chinese" "Chinese" "Chinese" "Chinese" "Chinese" "Malay"
 [8] "Malay"   "Malay"   "Malay"   "Malay"   "Malay"   "Indian"  "Indian"
[15] "Indian"  "Indian"  "Indian"  "Indian"  "Others"  "Others"  "Others"
[22] "Others"  "Others"  "Others"
> class(Race)
[1] "character"
> RaceFactor <- as.factor(Race)
> RaceFactor
 [1] Chinese Chinese Chinese Chinese Chinese Chinese Malay   Malay   Malay
[10] Malay   Malay   Malay   Indian  Indian  Indian  Indian  Indian  Indian
[19] Others  Others  Others  Others  Others  Others
Levels: Chinese Indian Malay Others
> class(RaceFactor)
[1] "factor"
```

For additional data type, data structure, R commands and syntax, one could refer to Lander (2014) and Teetor (2011).

1.8   Read File into R

Data are usually stored in different formats such as in a text file, a csv file, or a specified software of SPSS, SAS, STATA, and EXCEL format. There is no one unique and straightforward function, but rather using different functions, to read the different file formats into R environment. This section illustrates a few functions that are available in base and add-on packages to read in different file formats into R.

R function read.table() is to read in text file. The option header = TRUE specifies the heading, the first line of the text file, is read in as variable names. The data are specified after the "text=' " with one line per record. The output from the read.table () function is a data frame.

```
# ------------------- #
# Read in a Data Frame #
# ------------------- #
DF1 <- read.table(header=TRUE, text='
 T1 T2 T3 T4 T5 T6 T7 T8
A 1 1 2 3 4 5 6 NA
B 1 2 3 4 5 6 7 8
C 1 1 4 5 7 8 7 7
D 4 5 6 7 7 7 7 7
E 1 2 3 4 5 NA NA NA')
DF1
```

While data is stored in csv format, read.ccv() is the function to read this type of file format. A csv file is a text file containing data separated by commas, stored with a .csv extension. When opening a csv file, it usually invokes EXCEL to open it as if it is an EXCEL file. When the file is separated by semicolon (;) instead of comma (,), use read.csv2() function. The following shows syntax of reading in a csv file named as CSV1.csv, read into R, and stored as a data frame named as CSV1.

```
CSV1<- read.csv("CSV1.csv")
```

STATA, SPSS, and SAS are common statistical software with data store in their own file format. These three data formats could be read into R using functions in package foreign. Package foreign is an add-on package to read in STATA, SPSS, and SAS portable files. The functions for reading in these software files are read.dta(), read.spss(), and read.xport(), respectively. The following commands show the syntax of reading in STATA file STATA1.dta, SPSS file SPSS1.sav, and SAS portable file SAS1.xpt as data frame named as STATA1, SPSS1, and SAS1, respectively.

```
library(foreign)
STATA1 <- read.dta("C:/ ... /STATA1.dta")
SPSS1 <- read.spss("C:/ ... /SPSS1.sav")
SAS1 <- read.xport("C:/ ... /SAS1.xpt")
```

Microsoft EXCEL is another commonly used data storage format. Package xlsx is the package for reading in EXCEL file. Use read.xlsx() function to read in an EXCEL file. The syntax to read in an EXCEL is shown below.

```
library(xlsx)
EXCEL1 <- read.xlsx("C:/ ... /EXCEL1.xlsx", sheetName = "Sheet1")
```

Table 1.3 below summarizes the read files commands in base R and add-on package foreign and xlsx.

Exercises

1.1 Create the following vectors and calculate the following operations.

  (a) X + Y
  (b) Z + X
  (c) Z − Y

$$X = \begin{pmatrix} 1 \\ 2 \\ 3 \\ 4 \end{pmatrix}, Y = \begin{pmatrix} 2 \\ 2 \\ 2 \\ 2 \end{pmatrix}, Z = \begin{pmatrix} 4 \\ 3 \\ 2 \\ 1 \end{pmatrix}$$

1.2 Generate the following using R functions.

  (a) Create a sequence from −12 to 12.
  (b) Create a character factor contains two characters "Doubly Classified" and "Square Table."
  (c) Create a character factor that runs from "a" to "k."

Table 1.3 Functions to read files into R

Package/Function	Description
Base	
read.table()	Main function to read file into table format
read.csv()	Read csv file (file stored variable separated by comma ",")
read.csv2()	Read csv file separated by semicolon ";"
read.delim()	Read files separated by tabs "\t"
read.delim2()	Similar to read.delim()
read.fwf()	Read fixed width format files
Package foreign	
read.dta()	Read STATA file
read.spss()	Read SPSS file
read.xport()	Read portable SAS file
Package xlsx	
read.xlsx()	Read EXCEL file

(d) Create a character factor that runs from "A" to "Z."
(e) Put all the above into a list.

1.3 Create the following vector.
1, 2, 3, 4, 5, 10, 9, 8, 7, 6.

(a) Print the fourth element of the vector.
(b) Print the fifth–seventh element.
(c) Print those greater than 4.
(d) Print the first, third, and tenth element.

1.4 Create the following data frame.

A	B	C	D
1	3	TRUE	A
1	4	TRUE	B
1	8	TRUE	C
1	1	TRUE	D
1	3	TRUE	D
1	5	TRUE	F
2	7	FALSE	G
3	9	FALSE	D
4	3	FALSE	A
5	1	FALSE	B

i. Use r function to calculate the number of rows for this data frame.
ii. How many columns are there for this data frame?
iii. Use the subset command to carry out the following:

(a) Contain only 1 for A.
(b) Contain only TRUE for C.
(c) Contain only d for D.
(d) Contain only both A and B greater than 1.

References

Lander, J. (2014). *R for everyone: Advanced analytics and graphics* (2nd ed.). Addison-Wesley Data & Analytics Series.
Teetor, P. (2011). *R cookbook*. O'Reilly.

Chapter 2
Basic Concepts

2.1 Contingency, Square, and Doubly Classified Table

Contingency table, also known as cross-tabulation or crosstab, is a summary of data where the variables in classification are discrete factors or categorical variables, and the responses under the two crossing categorical variables are presented as counts. For a $n \times n$ contingency table, it is an output table format that allows for better visualization on the relationship between two categorical variables with n categorical coding for both row and column. Table 2.1 is a contingency table about the opinion of student's life satisfaction over a two year period. This contingency table is a 10×10 table with row representing student's life satisfaction level at year 1 and column representing year 2 life satisfaction. The codes for both column and row have the same value being 1 represents the lowest life satisfaction and 10 represents highest life satisfaction. There are nine students who expressed lowest life satisfaction and 65 expressed highest life satisfaction consistently over the 2 year period. Twenty students have an opinion of giving a 9 point in the first year but changed their opinion to a higher 10 point in the second year. This contingency table summarizes and tells us the changes in student's life satisfaction over the 2 year period. The number of observations stated in the table describes the association of the two time points of student's opinion about their life satisfaction. A contingency table is thus a type of table in a matrix format that displays the frequency distribution of the variables that help to understand how two categorical variables relate to each other.

Square Table
While contingency table describes the relationship of two categorical variables, the two variables could be entirely different in their codes and meanings. In general, a contingency table need not be $a \times a$, but any $n \times m$ table where n is differed from m. For instance, the row could be life satisfaction level of students with 10 codes and the column is student's race with 4 codes. A square table is a special type of contingency table that has same number of rows and columns. Table 2.1 is a square table with 10 rows and 10 columns, an $a \times a$ square table where $a = 10$.

© Springer Nature Singapore Pte Ltd. 2017
T.K. Tan, *Doubly Classified Model with R*,
https://doi.org/10.1007/978-981-10-6995-6_2

Table 2.1 Student's life satisfaction

		Year 2									
		1	2	3	4	5	6	7	8	9	10
Year 1	1	9	1	7	7	14	4	10	4	4	0
	2	5	6	6	5	6	3	3	3	4	2
	3	1	4	10	8	9	15	5	8	0	3
	4	7	6	8	10	35	25	15	12	7	2
	5	14	9	21	31	91	48	45	54	12	10
	6	4	2	10	13	46	54	80	34	11	3
	7	3	3	8	17	62	70	117	106	23	21
	8	4	4	10	18	46	52	126	141	60	19
	9	0	2	5	5	20	19	38	82	57	20
	10	4	1	6	8	18	22	31	48	43	65

While a square table requires same number of rows and columns, an additional qualification is needed to qualify as a doubly classified table. Doubly classified tables are tables with commensurable classification variables. That is, their codes are exactly the same with identical meanings. Table 2.1 is a square table as it has the same number of rows and columns. It is also a doubly classified table as the meanings attached to the codes for both row and column are conceptually with the same interpretation. Simply, doubly classified table is a special case of square table that their codes and the attached meanings are exactly the same for both the row and column. Hagenaars (1990) calls doubly classified table as turnover table, and Lawal (2003) refers it as doubly classified categorical data. Some authors do not distinguish between square table and doubly classified table and use them interchangeably. However, I shall refer and restrict to the cross-tabulations that have codes with identical property of a square table as doubly classified table.

2.2 Generating Frequency Tables

How to generate a contingency table in R? As data can be stored in different forms and formats and it may not necessarily in a cross-tabulation format, there are various ways to create a table. The following subsections illustrate with examples to show the various ways data can be read in and produced table using a number of libraries and functions.

2.2.1 Tables from Vector

The table() function is probably the most straightforward function to generate a table. This function can generate table from vector. The following shows two examples of generating two tables *TT* and *L* table from a string and a numeric vector, respectively. The table() function reads in the string vector Pets, counts the number of observations for each strings category and produces a table *TT*. The Likert is a numeric vector. Similarly, the table *L* table is produced by counting the number of observations of this numeric vector Likert.

```
Pets <- c('Dog','Cat','Duck','Chicken','Duck','Cat','Dog','Chicken')
TT <- table(Pets)
TT

    Pets
        Cat Chicken     Dog    Duck
          2       2       2       2

Likert <- c(1,1,1,1,1,1,2,2,2,2,2,3,3,3,4,4,4,4)
LTable <- table(Likert)
LTable

    Likert
    1 2 3 4
    6 5 3 4
```

2.2.2   Tables from Data Frame

The table() also generates table by reading from data frame. The following shows two examples of generating tables; one reading from a data frame with numeric elements and the other from a data frame with string elements.

```
AB <- data.frame(a=c(1,2,2,2,2,2,2,2,1,1),
                 b=c(2,1,1,1,2,1,2,1,2,1))
table(AB)

      b
    a   1 2
      1 1 2
      2 5 2

S <- data.frame(s1=c('A','A','A','B','B','B','C','C'),
                s2=c('A','A','B','B','B','C','C','C'))
table(S)

      s2
    s1  A B C
      A 2 1 0
      B 0 2 1
      C 0 0 2
```

Another way of producing table is to read indirectly as a data frame using read. table() function, followed by the table() function. The read.table() function below reads in two variables Year 2014 and Year 2015 with 10 observations into a data frame Data 1. The table() function tabulates these two variables into a table.

```
Data1 <- read.table(header=T,text='
Year2014 Year2015
     1          1
     1          1
     1          1
     2          1
     2          1
     1          2
     2          2
     2          2
     2          2
     2          2
')
TableData1 <- table(Data1$Year2014, Data1$Year2015)
TableData1
```

```
> TableData1

   1 2
 1 3 1
 2 2 4
```

2.2.3   Margin Total and Proportion

For any cross-tabulation, frequency is not the only statistics we are interested in; the total number of observations, margin total of rows and columns, row percentages, and column percentages are useful information to tell us about the association of two variables. The table() function gives only the frequencies. The margin.table() function generates the marginal frequencies, and the prop.table() function generates the proportions.

The function margin.table() outputs the margin total of a table. Without any option specified, the function returns the total number of observations of a table. With the specification of option 1, margin.table(TableData1,1) produces the row totals, giving the values of 4 and 6. With the specification of option 2, margin.table (TableData1,2) produces the column totals of 5 and 5. The command margin.table (TableData1) produces the total number of the observations of 10 observations. With the specification of option 1, margin.table(TableData1,1), it produces the row column totals of 2 and 6, and the specification of option 2, margin.table (TableData1,2), produces the column totals of 2 and 5.

```
> margin.table(TableData1)
[1] 10
> margin.table(TableData1, 1)

1 2
4 6
> margin.table(TableData1, 2)

1 2
5 5
```

Table 2.2 Summary of table functions and options

Function	Option	Description
table	–	Table of observations
margin.table	Default	Total number of observations
	(,1)	Row total
	(,2)	Column total
prop.table	Default	Cell %
	(,1)	Row %
	(,2)	Column %

The prop.table() function generates the cell percentages of a table. Option 1 and 2 generate the row and column percentages of a table, respectively.

```
> prop.table(TableData1)    # cell percentages

    1   2
1 0.3 0.1
2 0.2 0.4
> prop.table(TableData1, 1) # row percentages

          1         2
1 0.7500000 0.2500000
2 0.3333333 0.6666667
> prop.table(TableData1, 2) # column percentages

    1   2
1 0.6 0.2
2 0.4 0.8
```

Table 2.2 summarizes the three table functions discussed so far to generate table, marginal, and proportion.

2.2.4 xtab Function

The table() function is not the only function in R that generates table. The xtab() function is an alternative that is commonly used for tabulation of table. The syntax of xtab() is different from that of table(). The variables for tabulation are specified after the ~ sign. The + sign specifies the cross-tabulation of the variables before and after it. The data option specifies the data to read in. The following specification produces the cross-tabulation of Year 2014 and Year 2015 from the data frame Data 1.

TableData1xtab <- xtabs(~ Year2014 + Year2015, data=Data1)
TableData1xtab

```
> TableData1xtab
        Year2015
Year2014 1 2
       1 3 1
       2 2 4
```

2.2.5 Package gmodels, CrossTable Function

SAS users who are familiar with the PROC FREQ output format produced by SAS may find package gmodels suits them. Package gmodels, function CrossTable() produces the SAS format output as shown below. There are 5 rows of information produced by the CrossTable() function specified in the box in 5 rows as N, Chi-Square Contribution, Row % Total, Col % Total, and Table % Total. The cross-tabulation of Data1$Year2014 and Data$Year2015 produced by the CrossTable() shows the printout according to the row order printed in the box. For instance, for the cell(1,1), $N = 3$, Chi-Square Contribution = 0.5, Row % Total = 0.75, Col % Total = 0.60, and Table % Total = 0.3.

```
library(gmodels)
CrossTable(Data1$Year2014,Data1$Year2015)
```

```
   Cell Contents
|-------------------------|
|                       N |
| Chi-square contribution |
|            N / Row Total |
|            N / Col Total |
|          N / Table Total |
|-------------------------|

Total Observations in Table:  10

              | Data1$Year2015
Data1$Year2014 |        1 |        2 | Row Total |
---------------|----------|----------|-----------|
            1 |        3 |        1 |         4 |
              |    0.500 |    0.500 |           |
              |    0.750 |    0.250 |     0.400 |
              |    0.600 |    0.200 |           |
              |    0.300 |    0.100 |           |
---------------|----------|----------|-----------|
            2 |        2 |        4 |         6 |
              |    0.333 |    0.333 |           |
              |    0.333 |    0.667 |     0.600 |
              |    0.400 |    0.800 |           |
              |    0.200 |    0.400 |           |
---------------|----------|----------|-----------|
   Column Total |        5 |        5 |        10 |
              |    0.500 |    0.500 |           |
---------------|----------|----------|-----------|
```

2.2.6 Package descr, CrossTable Function

Similar to package gmodels, package descr, function CrossTable also produces tabulation with frequencies, cell percentage, row percentage, column total, row total, and overall total. The R syntax is given below.

```
library(descr)
CrossTable(Data1$Year2014,Data1$Year2015)
```

Table 2.3 Summary of commands for tabulation

Package	Purpose	R syntax
Base	Single table tabulation	table(A)
		xtabs($\sim$ A)
	Cross-tabulation: row A and column B	table(A, B)
		xtabs $\sim$ (A + B)
	A frequencies summed over B	T <- table(A, B); margin.table(T,1)
	B frequencies summed over A	T <- table(A, B); margin.table(T,2)
	Cell percentage	T <- table(A, B); prop.table(T)
	Row percentage (1 = rows)	T <- table(A, B); prop.table(T,1)
	Column percentage (2 = columns)	T <- table(A, B); prop.table(T,2)
	Include missing in table	table(A, B, exclude=NULL)
	3-way table	T <- table(A, B, C)
gmodels	1-way tabulation	CrossTable(A)
	2-way tabulation	CrossTable(A, B)
descr	1-way tabulation	CrossTable(A)
	2-way tabulation	CrossTable(A, B)

```
   Cell Contents
|-------------------------|
|                       N |
|   Chi-square contribution |
|           N / Row Total |
|           N / Col Total |
|         N / Table Total |
|-------------------------|


=====================================
                  Data1$Year2015
Data1$Year2014      1        2    Total
-------------------------------------
1                   3        1       4
                0.500    0.500
                0.750    0.250    0.400
                  0.6      0.2
                  0.3      0.1
-------------------------------------
2                   2        4       6
                0.333    0.333
                0.333    0.667    0.600
                  0.4      0.8
                  0.2      0.4
-------------------------------------
Total               5        5      10
                  0.5      0.5
=====================================
```

Table 2.3 summarizes the commands for tabulation for the base and the two add-on packages so far discussed.

2.3 Graphics for Tabulation

Tabulation can be reproduced in graphical output. This section illustrates the graphical output using bar chart to display the frequencies of a table. Doubly classified models in graphical forms will be discussed in Chap. 8.

There are a quite a number of graphical packages in R. The base package provides a variety of graphical functions. Package ggplots and lattice graphics are another two commonly used packages for graphics. These three packages are briefly discussed in the following three subsections.

2.3.1 Bar Plot—Base

Bar plot is one way to show the length of frequencies of a table into a chart. Table 2.4 shows the results from two surveys about the life satisfaction of 35 respondents.

The following syntax plots the above table into a bar plot using the barplot() function. The option main = specifies the title of the chart and prints it on top of the chart. The legend option specifies the values of the legend which are extracted from the variable RowCol of the TableJosS1 data frame. The ylab and xlab options specify the label of y-axis and x-axis, respectively. The col = rainbow(4) outputs the bar plot into four colors of rainbow. The colors selected are from the rainbow template, a pre-defined color scheme of rainbow color.

```
TableJobS1 <- read.table(header=T,text='
RowCol Freq
11 5
12 12
21 12
22 6')
barplot(TableJobS1$Freq, col=rainbow(4),
        main = "Table in Bar Chart",
        legend = TableJobS1$RowCol,
        ylab = "Frequency",
        xlab = "Table Cell")
```

From the bar chart, we can observe the symmetry of the frequencies of the four bars. The two diagonal cells (the first and the last bar) are almost equal in length. Similarly, the equal length of the two middle bars shows the same length of the two off-diagonal cells.

Table 2.4 Life satisfaction of survey 2016 and 2017

Year 2016	Year 2017	
	Satisfy	Not satisfy
Satisfy	5	12
Not satisfy	12	6

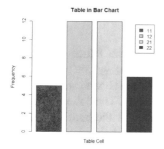

2.3.2   Package ggplot2

Similar bar plot could be carried out using package ggplot2, function ggplot. The aes option specifies the variable for the *x*-axis as RowCol and the *y*-axis as Freq and the color filled according to the number of categories of the variable RowCol. The geom_bar option specifies that the output is a bar chart. By default, geom_bar uses stat="bin". This command makes the height of each bar equal to the number of cases in each group. For the height of the bars to represent values in the data, the stat="identity" is specified to overwrite the default stat="bin". The show_guide option suppresses the legend by specifying it as FALSE. The theme_bw option takes away the frame of the graph. The ggtitle gives the title of the graph printed on the top of the graph.

```
library(ggplot2)
ggplot(data = TableJobS1, aes(x = RowCol, y = Freq, fill = RowCol)) +
      geom_bar(stat = "identity", show_guide = FALSE) +
      theme_bw() +
      ggtitle("Bar Chart")
```

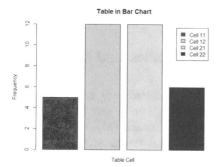

2.3.3 Package lattice

Another commonly used graphical package for R is the lattice package. The
function barchart() produces a bar chart similar to that of ggplot() function. The
main option specifies the title printed on top of the graph. The xlab and ylab options
specify the label for the *x*-axis and the *y*-axis, respectively. The col option specifies
the colors of the bar. As mentioned earlier, the rainbow is the built-in color specification to produce rainbow color output.

```
library(lattice)
barchart(Freq ~ RowCol,
         main="Bar Chart, Package lattice",
         xlab="Cell",
         ylab="Frequency",
         col=rainbow(4),
         data=TableJobS1)
```

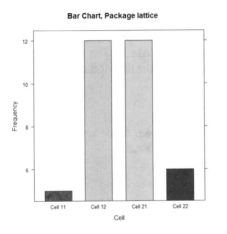

2.4 Odds, Odds Ratios, Local Odds Ratios, and Margin

The main aim of doubly classified modeling is to describe the association of cells,
summarized into an appropriate model or models that fit the data. There are various
measures to describe the association of cells. These include Odds, Odds ratios, local
Odds ratios, and marginal total. This section gives the basic concepts of these
measures.

2.4.1   Odds

There are several ways to describe a table. Frequency count, cell percentage, row percentage, column percentage, and marginal total are commonly used to describe association of cells for a table. Another way of donating percentages and probabilities is the Odds which is commonly used in doubly classified models.

The Odds of an event happening is the probability that the event will happen divided by the probability that the event will not happen. Given a probability π of success, the Odds, Ω, is defined as the probability of success, π, over the probability of failure, $1 - \pi$, as shown below.

$$\Omega = \frac{\pi}{1 - \pi}$$

For example, with probability $\pi = 0.75$, odds $\Omega = 0.75/0.25 = 3$. This means that a success is three times as likely as a failure. That is, we expect about three successes in every one failure. Inversely, when $\Omega = 1/3$, a failure is three times as likely as a success. The probability of success in terms of Odds is as follows:

$$\pi = \frac{\Omega}{1 + \Omega}$$

2.4.2   Properties and Interpretation of Odds

The odds are nonnegative values. When $\Omega = 1.0$, it indicates that the likelihood of success is equal to the likelihood of failure. With $\Omega > 1.0$, a success is more likely than a failure. That is, the event is more likely to happen than not. With $\Omega < 1.0$, a failure is more likely than a success. This means that the event is less likely to happen.

The following table shows the preference of Singapore residents whether they prefer to live in the east or west part of the island.

From the above table, we can see that there is about 61.5% of Singapore residents preferred to live in the east and 38.5% preferred to live in the west. The odds that a Singapore resident preferred living at the east as opposed to the west are about 1.6 (5600/3500). That is, living in the east is 1.6 more likelihood than living in the west part of Singapore.

$$\text{East } \% = \frac{5600}{9100} = 61.5\% \quad \text{West } \% = \frac{3500}{9100} = 38.5\%$$

$$\text{odds} = \frac{5600}{3500} = \frac{0.615}{1 - 0.615} = 1.6$$

The R syntax to calculate the odds is as follows:

```
East <- 5600
West <- 3500
Odds <- East / West
Odds
```

Alternatively, odds can be derived by dividing two probabilities. Both PEast and PWest are expressed in probability term. The odds is the probability of the preferences to live in east over west.

```
PEast <- East / (East+West)
PWest <- West / (East+West)
Odds <- PEast / PWest
Odds
```

Table 2.6 further breakdowns the residence preference of Table 2.5 into current and past. Here, we want to know whether there is association between current and past preferred residence and the difference of their odds for those residents is living in the east.

Let the probability of success π_1 for the east residence (row 1) living in the east, and π_2 for east residence living in the west (row 2). The odds of current east residences in favor of living at the east in the past is $Odds_1 = 3.6$ (3600/1000), whereas the odds of current east residence in favor of living at the west in the past is $Odds_2 = 0.8$ (2000/2500). The odds of living in the same area (3.6) is higher than change in area (0.8). Similarly, we can calculate the odds of current west residences in favor of living at the east in the past has an odds of 0.28 (1000/3600) and the odds of current west residence in favor of living at the west in the past is 1.25 (2500/2000). Likewise, the odds of living in the same area (1.25) is higher than change in area (0.28). These results show that there is a tendency of preference residence living in the same part of the island. There is an association between current residence and preference. The east residences prefer to live in the east, whereas the west residences prefer to live in the west.

Table 2.5 Residential preference

Preference	
East	West
5600	3500

Table 2.6 Current residence × Past residential preference

Past residence preference	Current preference	
	East	West
East	3600	1000
West	2000	2500
Total	5600	3500

2.4.3 Odds Ratio

Odds ratio is another way to describe the relationship of cells. As the name implies, it quantifies the strength of association by computing the ratio of two odds. The following formula defines odds ratio, θ, as an odds Ω_1 over another odds Ω_2.

$$\text{Odds Ratio} = \theta = \frac{\Omega_1}{\Omega_2} = \frac{\text{Odds}_1}{\text{Odds}_2} = \frac{\frac{\pi_1}{1-\pi_1}}{\frac{\pi_2}{1-\pi_2}}$$

For a 2×2 table, the odds ratio is the ratio of two odds. Table 2.7 is a 2×2 table with four cells of frequency n_{11}, n_{22}, n_{12}, and n_{21}. The odds ratio is simply the ratio of taking the first row odds n_{11}/n_{12} over the second row odds n_{21}/n_{22}. Sometimes, it is also referred to as the cross product because the odds ratio is the product of the diagonal elements $(n_{11} \times n_{22})$ by the product of the off-diagonal elements $(n_{12} \times n_{21})$:

$$\theta = \frac{n_{11}/n_{12}}{n_{21}/n_{22}} = \frac{n_{11}n_{22}}{n_{12}n_{21}}$$

The sample odds ratio could also be viewed in probability terms where $p_1 = n_{11}/n_{11} + n_{12}$ and $p_2 = n_{21}/n_{21} + n_{22}$. The odds ratio is the odds $p_1/(1 - p_1)$ over the odds $p_2/(1 - p_2)$. The following shows by changing the probability terms into frequencies; it becomes the cross product term $n_{11}n_{22}/n_{12}n_{21}$. For cross-tabulation, applying the formula using cross product frequency is more direct way to calculate odds ratio.

$$\hat{\theta} = \frac{p_1/(1-p_1)}{p_2/(1-p_2)} = \frac{\frac{n_{11}/n_{11}+n_{12}}{1-n_{11}/n_{11}+n_{12}}}{\frac{n_{21}/n_{21}+n_{22}}{1-n_{21}/n_{21}+n_{22}}} = \frac{n_{11}/n_{11}+n_{12}-n_{11}}{n_{21}/n_{21}+n_{22}-n_{21}} = \frac{n_{11}n_{22}}{n_{12}n_{21}}$$

A value of 1 for an odds ratio implies there is no association between the variables. The farther away from the value of 1 of an odds ratio, the stronger is the association. The values of 4 and 0.25 are equally distant from 1 since $0.25 = \frac{1}{4}$. A value and its reciprocal refer not just to the same strength of association but also the same association, in terms of odds.

Table 2.7 2×2 Table

A	B	
	1	2
1	n_{11}	n_{12}
2	n_{21}	n_{22}

It is very likely in practice we may come across frequency cells that contain zero in a table, in particularly when the table is large. Using the above formula, the odds ratio will turn out as either a zero or an infinite. If n_{11} or $n_{22} = 0$, the estimated odds ratio turns out as zero, and if n_{12} or $n_{21} = 0$, the odds ratio produces infinity. This is not an ideal way of calculating and representing odds ratio. In order to avoid this bizarre outcome, an adjusted odds ratio is normally used. This is usually carried out by adding a small number of 0.5 to all the cell frequencies, as shown below.

$$\hat{\theta} = \frac{(n_{11}+0.5)(n_{22}+0.5)}{(n_{12}+0.5)(n_{21}+0.5)}$$

For a 2×2 table, there is only one odds ratio, but in larger tables, there are more than one odds ratio. In general, for a $I \times J$ table, there are $(I-1)(J-1)$ adjacent odds ratios. We refer to these odds ratios as the local odds ratios, defined as below.

$$\hat{\theta}_{ij} = \frac{n_{ij}n_{(i+1)(j+1)}}{n_{(i+1)j}n_{i(j+1)}}$$

Table 2.8 shows a 4×4 table with 16 cells. Table 2.9 produces a 3×3 adjacent odds ratios table with nine cells. For instance, $\theta_{11} = n_{11}n_{22}/n_{12}n_{21}$, and $\theta_{13} = n_{13}n_{24}/n_{14}n_{23}$. In general, an $a \times a$ table will produce an a-1 $\times$ a-1 odd ratios table.

Table 2.8 4×4 Table

A	B			
	1	2	3	4
1	n_{11}	n_{12}	n_{13}	n_{14}
2	n_{21}	n_{22}	n_{23}	n_{24}
3	n_{31}	n_{32}	n_{33}	n_{34}
4	n_{41}	n_{42}	n_{43}	n_{44}

Table 2.9 4×4 Odds ratios table

A	B		
	1	2	3
1	θ_{11}	θ_{12}	θ_{13}
2	θ_{21}	θ_{22}	θ_{23}
3	θ_{31}	θ_{32}	θ_{33}

Table 2.10 Importance of religious belief of husband and wife

Wife importance of religious belief	Husband importance of religious belief		
	Not	Fairly	Very
Not important	56	32	39
Fairly important	43	61	37
Very important	38	20	20

Table 2.11 Local odds ratios

Wife importance of religious belief	Husband importance of religious belief	
	Not versus fairly	Fairly versus very
Not versus fairly	2.48	0.50
Fairly versus very	0.37	1.65

The following is a 3×3 cross-tabulation of husband and wife belief about the importance of religions.

The local odds ratios of adjacent cells are calculated as follows, and these odds ratios are summarized in Table 2.11.

$$\hat{\theta}_{11} = \frac{n_{11}n_{22}}{n_{21}n_{12}} = \frac{56 \times 61}{43 \times 32} = 2.48 \quad \hat{\theta}_{12} = \frac{n_{12}n_{23}}{n_{22}n_{13}} = \frac{32 \times 37}{61 \times 39} = 0.50$$

$$\hat{\theta}_{21} = \frac{n_{21}n_{32}}{n_{31}n_{22}} = \frac{43 \times 20}{38 \times 61} = 0.37 \quad \hat{\theta}_{22} = \frac{n_{22}n_{33}}{n_{32}n_{23}} = \frac{61 \times 20}{20 \times 37} = 1.65$$

For a 3×3 table, there are four adjacent odds ratios. However, odds ratios are not restricted to adjacent cells. There are local odds ratios not referring to adjacent cells. For instance, we could take cell (1,1), cell (3,3), cell (1,3), and cell (3,1) to derive an odds ratio.

$$\hat{\theta} = \frac{n_{11}n_{33}}{n_{31}n_{13}} = \frac{56 \times 20}{38 \times 39} = 0.76$$

2.4.4 Log Odds Ratio and Confidence Interval

Another measure that often used in doubly classified modeling is log odds ratio. Log odds ratio, Φ_{ij}, is simply taking logarithm of odds ratio.

$$\Phi_{ij} = \ln\left(\theta_{ij}\right)$$

As odds ratio ranges from zero to infinity, the sampling distribution of the odds ratio is positively skewed to the right. Unlike odds ratio, log odds ratio is approximately normally distributed with a bell-shaped. This makes log odds ratio useful for construction of confidence interval for odds ratio.

The confidence interval is normally computed to indicate the level of uncertainty around the measure of odds ratio. As study usually recruits only a small sample of the overall population, the precision of the odds ratio, expressed in confidence intervals, specifies an upper and lower confidence limit that infers the true population effect lies between these two points. Most studies report the 95% confidence interval by specifying α as 0.05 and $z_{\alpha/2}$ as 1.96. The asymptotic standard error (ASE) of the log odds ratio is given below.

$$ASE\left(\log \hat{\theta}\right) = \sqrt{\frac{1}{n_{11}} + \frac{1}{n_{12}} + \frac{1}{n_{12}} + \frac{1}{n_{22}}}$$

The $(1-\alpha)\%$ confidence interval for log odds ratio is

$$\log \hat{\theta} \pm z_{\alpha/2} ASE\left(\log \hat{\theta}\right)$$

and the $(1-\alpha)\%$ confidence interval for odds ratio is

$$\left(e^{\log \hat{\theta} - z_{\alpha/2} ASE}, e^{\log \hat{\theta} + z_{\alpha/2} ASE}\right)$$

The procedure to calculate the confidence interval of odds ratio is as follows:

(1) Calculate odds ratio
(2) Calculate the natural log of the odds ratio
(3) Use the above formula to calculate confidence coefficient using standard normal distribution
(4) Take exponential to convert the upper and lower log odds ratio back to odds ratio

The following R syntax shows the calculation of odds ratio for Table 2.6.

```
Residence <- matrix(c(3600, 2000, 1000, 2500), ncol=2)
OR <- (Residence[1,1] * Residence[2,2]) / (Residence[1,2] * Residence
[2,1])
OR
```

```
> OR
[1] 4.5
```

$$\text{Odds Ratio} = \frac{3600 \times 2500}{1000 \times 2000} = 4.5$$

Below is a function called odds.ratio written to calculate odds ratio, asymptotical standard error, and confidence interval of odds ratio.

```
odds.ratio <- function(x, zeros=FALSE, conf.level=0.95) {
    if (zeros) {
        if (any(x==0)) x <- x + 0.5
    }
    OR <- x[1,1] * x[2,2] / ( x[2,1] * x[1,2] )
    ASE <- sqrt(sum(1/x))
    CI <- exp(log(OR) + c(-1,1) * qnorm(0.5*(1+conf.level)) *ASE )
    list(Odds.Ratio=OR,
        ASE=ASE,
```

```
        conf.interval=CI,
        conf.level=conf.level)
}
odds.ratio(Residence)
```

```
> odds.ratio(Residence)
$Odds.Ratio
[1] 4.5

$ASE
[1] 0.04666667

$conf.interval
[1] 4.106670 4.931003

$conf.level
[1] 0.95
```

The estimated odds ratio for the Residence is 4.5. The upper and lower confidence intervals for the odds ratio with 95% confidence interval are 4.11 and 4.93, respectively. Since the confidence intervals do not cover 1, it implies there is association between current and preferred residence.

Although the above function odds.ratio does well for odds ratio output, there are at least three add-on R packages that provide functions to generate odds ratio. They are tabulated in the following table.

The following shows the output from package fmsb on odds ratio estimates and 95% confidence intervals. It is noted that the heading of the table is stated as "Disease" and "Nondisease". This is to do with the way the package is written. It is developed for the purpose of medical statistics book with demographic data, so the output is not a general heading which the users could specify. Nonetheless, the output shows the same results of the written function odds.ratio.

```
library(fmsb)
oddsratio(3600,1000,2000,2500,conf.level=0.95)
```

```
> oddsratio(3600,1000,2000,2500,conf.level=0.95)
             Disease Nondisease Total
Exposed         3600      2000   5600
Nonexposed      1000      2500   3500
Total           4600      4500   9100

        Odds ratio estimate and its significance probability

data:   3600 1000 2000 2500
p-value < 2.2e-16
95 percent confidence interval:
 4.106670 4.931003
sample estimates:
[1] 4.5
```

Table 2.10 is reused here to illustrate the calculation of confidence interval of odds ratio using log odds ratio. The following table shows the log odds ratio of Table 2.11.

The general formula for log local odds ratio is as follows:

$$\log\left(\hat{\theta}_{ij}\right) = \log\left(n_{ij}\right) + \log\left(n_{(i+1)(j+1)}\right) - \log\left(n_{(i+1)j}\right) - \log\left(n_{i(j+1)}\right)$$

For large sample, the sampling distribution of the log odds ratio is approximately normal with the following specification of asymptotic standard error.

$$ASE\left(\log \hat{\theta}_{ij}\right) = \sqrt{\frac{1}{n_{ij}} + \frac{1}{n_{(i+1)j}} + \frac{1}{n_{i(j+1)}} + \frac{1}{n_{(i+1)(j+1)}}}$$

To calculate a 95% confidence interval for the local odds ratio between not, fairly, and very important for husband and wife, the standard error for the log odds ratio is as follows:

$$\sqrt{\frac{1}{56} + \frac{1}{43} + \frac{1}{32} + \frac{1}{61}} = \sqrt{0.0888} = 0.2979$$

The confidence interval for log odds ratio is thus

$$0.908 \pm 0.2979 \times 1.96 = (0.3241, 1.4919)$$

After taking the exponentiation, the confidence interval for odds ratio is

$$\left(e^{0.3241}, \ e^{1.4919}\right) = (1.3828, \ 4.4455)$$

The 95% confidence interval for the odds ratio in the population does not include one. We reject the hypothesis that there was no association between husband and wife of not and very important.

2.4.5 Margin Total

The concept of margin total is essential for doubly classified model as there are doubly classified models that are based on this property. For a 2×2 table, Table 2.12 marginal total are n_{1+}, n_{2+}, n_{+1}, and n_{+2}. The total margin for the table is n_{++} (Tables 2.13 and 2.14).

Table 2.12 R functions for odds ratio

Package	Function
epitools	oddsratio
fmsb	oddsratio
hypergea	getOddsRatio

Table 2.13 Local log odds ratios

Wife importance of religious belief	Husband importance of religious belief	
	Not versus fairly	Fairly versus very
Not versus fairly	0.908	−0.693
Fairly versus very	−0.994	0.501

Table 2.14 2×2 Table, margin total

A	B		Total
	1	2	
1	n_{11}	n_{12}	n_{1+}
2	n_{21}	n_{22}	n_{2+}
Total	n_{+1}	n_{+2}	n_{++}

2.5 Applications of Doubly Classified Data

There are numerous situations where doubly classified table is applicable and useful for analysis. This section groups the applications of doubly classified table into five main areas listed below.

(1) Matched pairs
(2) Essential similar variables
(3) Longitudinal study for a common variable
(4) Inter-rater agreement
(5) Indicators from a scale

2.5.1 Studies on Pairs of Matched Individuals

Studies on matched pairs are common. Examining twins, husband and wife, father and son are examples of matched pairs. Study on husband and wife opinion on bringing up their children is an application of matched pairs. Comparison of occupational status of father and son pairs is a typical example of using matched pairs to study intergeneration mobility pattern. Study designs such as crossover clinical trials, matched cohort studies, and matched case-control studies, pre- and post-intervention programs are designs using matched pairs.

Analyzing pair of matched individuals by putting them into a table is one common way to find out paired relationships. A social mobility table describes the mobility pattern in the form of a table to find out intergenerational social mobility. The columns and the rows have the same descriptor to specify socioeconomic position of two generations, and the frequencies tell the story about the mobility pattern between son and their parental socioeconomic position. The following

Table 2.15 Father's and son's occupation status

Father's status	Son's occupation status					Total
	(1)	(2)	(3)	(4)	(5)	
(1)	50	45	8	18	8	129
(2)	28	174	84	154	55	495
(3)	11	78	110	223	96	518
(4)	14	150	185	714	447	1510
(5)	3	42	72	320	411	848
Total	106	489	459	1429	1017	3500

Table 2.15 is a typical social mobility table, taken from Hauser (1978) and Agresti (1983), which describes father's and son's occupation mobility of Britain in 1949. The number (1) in the table represents professional and high administrative; (2) represents managerial and executive; inspectional, supervisory, and other non-manual (higher grade); (3) represents inspectional, supervisory, and other non-manual (lower grade); (4) represents skilled manual and routine grades of non-manual; and (5) represents semi-skilled manual and unskilled manual. The order of these numbers shows the hierarchy of occupation being low in ranking of (5) to high ranking of (1). The diagonal cells represent those son and father occupation status that matched. Father with higher occupational status than son are those cells located at the upper diagonal while the lower diagonal cells are those father with lower occupational status than son. The codes used in table are of ordered categories. That is, there is an order of the code from order to high being (1) is the highest and (5) is the lowest. In general, mobility table can be placed as an $a \times a$ contingency table in which both rows and columns are ordered by social status.

Income mobility is another example that shows the association between parent's and child's income in a contingency square table. These mobility tables are typical pairs of individuals matched that are suitable for using doubly classified models to understand the pattern within the table. Table 2.12 is another example of a social mobility table (Table 2.16).

Table 2.17 is another doubly classified table that describes the association between husband and wife about intermarriage. It records the ethnicity of husband and wife married in the USA (Smith et al. 1996).

2.5.2 Association Between Two Essentially Similar Variables

Another application of doubly classified table is to examine association between two essentially similar variables. The strength of right and left hand, right and left eye grade for the same person are two examples that use a common measure to relate two essentially similar variables. In general, studies that examining a subject

Table 2.16 Father's and son's occupation status

Father's occupation status	Son's occupation status				
	Professional and high administrative	Managerial, executive, and high supervisory	Low inspectional and supervisory	Routine non-manual and skilled manual	Semi- and unskilled manual
Professional and high administrative	18	17	16	4	2
Managerial, executive, and high supervisory	24	105	109	59	21
Low inspectional and supervisory	23	84	289	217	95
Routine non-manual and skilled manual	8	49	175	348	198
Semi- and unskilled manual	6	8	69	201	246

Table 2.17 Husband's ethnicity by wife's ethnicity for immigrants married in the USA

Husband's ethnicity	Wife's ethnicity							
	British	Irish	Scandinavian	German	Italian	Polish	European Jewish Central	European Jewish Eastern
British	314	63	10	15	0	1	1	0
Irish	27	625	2	5	0	0	0	0
Scandinavian	4	9	835	20	1	0	0	0
German	26	26	10	1096	0	4	0	0
Italian	3	6	0	4	477	1	0	0
Polish	1	0	0	7	0	421	0	0
European Jewish Central	1	0	0	1	0	1	112	11
European Jewish Eastern	1	0	0	1	0	1	30	347

Table 2.18 Unaided distance vision of right and left eye

Right eye	Left eye				Total
	1	2	3	4	
1 Highest grade	152	27	12	7	198
2	23	151	43	8	225
3	12	36	177	20	245
4 Lowest grade	4	8	18	49	79
Total	191	222	250	84	747

responses to two similar situations, two similar instruments, and two similar states of mind are examples of two essentially similar variables that are suitable to put them in a doubly classified table to find out the relationships of the two variables.

The following table tabulates the unaided distant vision of 747 women. The ordered categories record a highest score of (1) to a lowest score of (4) in their visions, tabulated in Table 2.18. Here, we are interested to know whether a women's right eye grade is associated with her left eye grade and to know whether both eyes are in symmetry or asymmetry in their visions.

2.5.3   Two Point Longitudinal Study for a Common Variable

Doubly classified table also suits well for longitudinal studies in examining the changes over two point time for a common measure. When a common measure is used for the same person responded over two time period, a doubly classified table displays the changes over the two point time. In the area of medicine, psychology, and sociology studies when the variable of interest is measured before and after a treatment, doubly classified table is suitable to examine whether there is a change, and if there is, the pattern of change. Table 2.19 is a cross-tabulation about the voting transitions of two British surveys on election carried out at two different point in time, year 1966 and 1970 (Upton 1978). It is a constituency contested by the conservative, labor, and liberal parties. As such the electoral affiliations are categorized into four main groups, a symmetric classification to form a doubly classified table.

Table 2.20 shows another use of doubly classified table over two time points. The table shows the mobility movements between religious groups in Northern Ireland, extracted from International Social Survey Program's (ISSP) religion survey, 1991 (Breen and Hayes 1996). Social scientists are interested in whether

Table 2.19 British election study

Year 1966	Year 1970			
	Conservative	Liberal	Labor	Abstention
Conservative	68	1	1	7
Liberal	12	60	5	10
Labor	12	3	13	2
Abstention	8	2	3	6

Table 2.20 Origin and current religion—Northern Ireland

Origin religion	Current religion					
	Catholic	Anglican	Mainline Protestant	Fundamentalist Protestant	Other Protestant	None
Catholic	266	1	1	0	1	13
Anglican	5	137	24	5	3	20
Mainline Protestant	2	23	213	9	9	22
Fundamentalist Protestant	0	3	2	11	2	4
Other Protestant	0	2	2	1	7	1
None	0	0	1	1	0	7

there is an increasing volatility of religious affiliation over time. The secularization theorists believe that there is a gradual erosion of all religious identities starting with the more conservative religious denominations, whereas the rational choice theorists argue that the declination of conservative religious is due to its obsolescent on the quality of their recruits and restrictiveness placed on members. Using a two point data to tabulate a religious mobility table provides the changes of religious over time and their movements into a doubly classified table. The doubly classified model could then be carried out to examine the structural changes of religious and religious pluralization.

2.5.4   Inter-rater Agreement

Establishing rater agreement is usually carried out in fieldwork research to ensure raters apply the rubrics consistently in their assessments. If two raters coded consistently, a doubly classified table will show off-diagonal cells with zero observations, and if they fail to be 100% consistent in their assessment, the patterns of rater disagreement could be identified using doubly classified models.

The following two tables tabulate the assessment of rater A, B, C, and D. Table 2.21 gives the assessment rating of rater A and rater B, and Table 2.22 tabulates the results of rater C and rater D. It is observed that the rating of rater

Table 2.21 Inter-rater agreement of assessment

Rater B	Rater A			
	Grade 1	Grade 2	Grade 3	Grade 4
Grade 1	68	1	1	1
Grade 2	12	120	5	11
Grade 3	15	31	110	2
Grade 4	8	2	3	102

Table 2.22 Inter-rater
agreement of assessment

Rater D	Rater C			
	Grade 1	Grade 2	Grade 3	Grade 4
Grade 1	12	21	11	31
Grade 2	12	12	5	61
Grade 3	15	31	16	12
Grade 4	89	21	32	11

A and B are much more consistent than the rating between rater C and D. The percentage of main diagonal cells to the total observations of Table 2.21 (400/492 = 81.3%) is much higher than that of Table 2.22 (51/392 = 13%). Although it is essential to note that the percentage of agreement of rater A and B is much higher than rater C and D, doubly classified model can go beyond to examine and find out whether there is any specific pattern in their agreement and disagreement.

2.5.5 Two Indicators from a Scale

Another application of doubly classified model is to use it for examining the relationship of indicators when constructing a scale. In social science studies, using indicators to form a scale to represent a construct is a common practice. Within a scale, usually there are several indicators to represent the different aspects of the scale. These indicators normally have a common code which is appropriate for doubly classified modeling to understand their associations. Table 2.23 shows a doubly classified table of two indicators, literacy and leadership skills (Tan and Sheng 2015). The relationship of the two skills can be modeled using doubly classified models.

Table 2.23 Literacy and leadership skills

Literacy	Leadership									
	1	2	3	4	5	6	7	8	9	10
1	31	8	21	30	26	26	48	20	20	42
2	2	4	4	6	1	3	6	0	4	7
3	3	3	5	6	1	9	10	6	7	5
4	5	4	3	8	17	31	36	14	9	23
5	0	3	1	11	13	5	9	6	7	9
6	5	1	5	17	7	14	27	13	8	16
7	6	1	2	26	26	35	76	22	29	37
8	2	1	5	16	10	14	26	13	16	16
9	1	0	1	15	15	26	44	29	30	33
10	3	2	4	19	14	31	115	57	80	203

Exercises

2.1 Create the following matrix and convert it into a table and a factor.

$$\begin{pmatrix} 1 & 2 & 3 & 4 \\ 2 & 3 & 4 & 5 \\ 3 & 4 & 5 & 6 \\ 8 & 9 & 9 & 8 \end{pmatrix}$$

2.2 Create the following 2 tables using R functions (Tables 2.24 and 2.25).

2.3 Read in data infert from the base. This data have the following variables (Table 2.26).

    a. Show the structure of the data infert.

    b. List the first 6 and last 6 observations.

    c. Print the summary statistics using summary() function.

    d. Tabulate table frequency of induce × spontaneous.

    e. Add column and row margin to the table.

Table 2.24 Watching

Shopping with children	Watching TV with children		
	Maybe	No	Yes
Always	2	0	0
Never	0	2	1
Sometimes	2	1	1

Table 2.25 Gender × Smoking

	Smoke	No smoke
Male	700	120
Female	665	140

Table 2.26 Variable, description, and coding

Variable	Description	Coding
Education	Education (years)	0 = 0–5 years 1 = 6–11 years 2 = 12+years
Age	Age in years of case	21–44
Parity	Count	1–6
Induced	Number of prior induced abortions	0,1,2=2 or more
Case	Case status	1 = case; 0 = control
Spontaneous	Number of prior spontaneous abortions	0,1,2=2 or more
Stratum	Matched set number	1–83
Pooled.stratum	Stratum number	1–63

f. Add only column margin to the table.
g. Add only row margin to the table.
h. Generate a table with cell percentage. Restrict to 2 decimal places.
i. Generate a table with row cell percentage. Restrict to 2 decimal places.
j. Generate a table with column cell percentage. Restrict to 2 decimal places.
k. Add margin total to row cell percentage table.
l. Add margin total to column cell percentage (Table 2.4).

2.4 Read in data mtcar from the base. The data were extracted from the 1974 Motor Trend US magazine and comprise fuel consumption and 10 aspects of automobile design and performance for 32 automobiles (1973–1974 models). Carry out the following.

a. Show the structure of the data mtcar.
b. List the first 6 and last 6 observations.
c. Print the summary statistics using summary() function.
d. Print the frequency of variable vs.
e. Print the frequency of variable am.
f. Tabulate table frequency of vs × am.
g. Calculate the margin totals of the row for table vs × am.
h. Calculate the margin totals of the column for table vs × am.
i. Tabulate the cell percentage of table vs × am.
j. Tabulate the row percentage of table vs × am.
k. Tabulate the column percentage of table vs × am.
l. Use CrossTable function from Package gmodels to generate table vs × am.

2.5 Calculate the odds of region of origin, odds ratios for the (Table 2.27).
2.6 Calculate the odds of region of origin for the following table (Table 2.28).

Table 2.27 Current Residence × Residential Preference

Region of origin	Preference	
	North	South
North	3092	958
South	959	3027
Total	4051	3985

Table 2.28 Preferred camp location

Race	Preference	
	Camp A	Camp B
Chinese	2027	2268
Malay	2024	1717
Total	4051	3985

Table 2.29 p, q, and odds

p	q	Odds
0.8		
0.6667		
0.50		
0.40		
0.3333		
0.25		
0.20		

Table 2.30 Variable X and Y

Table 2.30.1				Table 2.30.2		
T1	Y			T2	Y	
X	1	2		X	1	2
1	20	20		1	200	200
2	10	10		2	100	100
Table 2.30.3				Table 2.30.4		
T3	Y			T4	Y	
X	1	2		X	1	2
1	20	10		1	10	20
2	10	20		2	20	10

2.7   Calculate the odds of the following p and interpret the results (Table 2.29).

2.8   Generate a vector with sequence from 0 to 1.0 with increment of 0.01 and calculate the odds, log of odds for this sequence. Plot the sequence against odds and the sequence against log of odds separately. What can you observe from the two graphs?

2.9   Calculate the odds ratio and log odds ratio for the following four tables of variable X and Y and comment (Table 2.30).

2.10 Generate odds ratio distribution and log odds ratio distribution by specifying the total number of observations as 100. Plot the two distributions.

References

Agresti, A. (1983). A simple diagonals-parameter symmetry and quasi-symmetry model. *Statistics and Probability Letters, 1*(6), 313–316.

Breen, R., & Hayes, B. C. (1996). Religious mobility in the UK. *Journal of the Royal Statistical Society, 159*(3), 493–504.

Hagenaars J. A. (1990). *Categorical longitudinal data. Log-linear Panel, Trend, and Cohort Analysis*. Newbury Park: Sage.

Hauser, R. M. (1978). A structural model of the mobility table. *Social Forces, 65*(3), 919–953.

Lawal, H. B. (2003). *Categorical Data Analysis with SAS and SPSS Applications*. Lawrence Erlbaum Associates, Publishers.

Smith, P. W. F., Foster, J. J. & McDonald, J. W. (1996). Monte Carlo exact test for square contingency tables. *Journal of Royal Statistical society A, 159*, 309–321.

Stuart, A. (1953). The estimation and comparison of strength of association in contingency tables. *Biometrika, 40*, 105–110.

Tan, T. K. & Sheng, Y. Z. (2015). *Extending the quasi-symmetry model: Quasi-symmetry with n degree*. Poster presented at the useR! Conference 2015.

Upton, G. J. G. (1978). *The Analysis of Cross-tabulated Data*. New York: Wiley.

Chapter 3
Examining Symmetry of Doubly Classified Table

Doubly classified model aims to examine pattern of cells within a table. In particularly, we look for symmetrical pattern inside a table. There are several ways to look at symmetry. The following lists the possibilities of symmetrical patterns.

1. Symmetry in cell probability
2. Symmetry in group of cells probability
3. Symmetry in odds
4. Symmetry in group of odds
5. Symmetry in odds ratio
6. Symmetry in group of odds ratios
7. Symmetry to the center of table
8. Combinations of the above
9. Symmetry in marginal total

The above lists nine possible symmetrical patterns that can be modeled using doubly classified models. The first eight symmetrical patterns will be discussed in subsequent chapters. The ninth symmetrical pattern is the focus of this chapter. It states that marginal total is in symmetry. This means that for an a × a doubly classified table, marginal totals from row 1 to a is in symmetrical to column 1 to a, respectively. This property is referred to as marginal homogeneity. Marginal homogeneity in doubly classified table context means the row total is equal to the column total in probabilistic term. For instance, analyses to assess axial symmetry of a doubly classified table between two time points for a common variable are often carried out to examine whether variables move from one category of response to another over time is marginally homogenous. If the response is not random, we would expect some kinds of symmetry pattern exist between two time points.

Specifically, this chapter concentrates on McNemar's χ^2 test (1947) of symmetry for 2 × 2 doubly classified table, its variations, Bower's test of symmetry for a × a doubly classified table and ends up with a discussion of the marginal symmetry/homogeneity model. The main aim is to test for marginal homogeneity.

© Springer Nature Singapore Pte Ltd. 2017
T.K. Tan, *Doubly Classified Model with R*,
https://doi.org/10.1007/978-981-10-6995-6_3

3.1   McNemar's Test of Symmetry

McNemar's test of symmetry, named after Quinn McNemar who introduced it in 1947, aims to test whether the table is in symmetry on rows and columns for a two-dimensional contingency table. It is also referred to as the symmetry McNemar Chi-Square test as it uses Chi-Square statistics for testing the hypothesis. This test is suited for dichotomous variables with two dependent sample studies, aims to assess if a statistically significant change in proportions has occurred on a dichotomous trait at two time points on the same population.

Table 3.1 above shows a 2×2 contingency table with the number of observations recorded as a, b, c, and d for cell (1,1), cell (1,2), cell (2,1), and cell (2,2), respectively. McNemar's test of symmetry is about testing marginal homogeneity of whether the two marginal probabilities for each outcome are the same at population level. The null hypothesis is whether the probabilities of being classified into cells [1,2] and [2,1] are the same at the population level. Put it in another way, it aims to test whether the proportions of the two categories of two matched groups is the same. The marginal total of $\pi_a + \pi_b = \pi_a + \pi_c$ turns out as $\pi_b = \pi_c$, and similarly testing of $\pi_c + \pi_d = \pi_b + \pi_d$ is equivalent of $\pi_c = \pi_b$. As such, only π_b and π_c are relevant for testing of marginal homogeneity. The null hypothesis for McNemar's test of symmetry is thus stated below.

$$H_0 : \pi_b = \pi_c$$
$$H_1 : \pi_b \neq \pi_c$$

or, alternatively

$$H_0 : O_{12} = O_{21}$$
$$H_1 : O_{12} \neq O_{21}$$

where O_{12} and O_{21} are observed frequencies occurring outside the main diagonal of the matrix of 2×2 contingency table. The Chi-Square statistics for McNemar's test of marginal homogeneity is with 1 degree of freedom.

$$\chi^2 = \frac{(b-c)^2}{b+c}$$

$$df = (\text{rows} - 1)(\text{columns} - 1) = (2-1)(2-1) = 1$$

Table 3.1 2×2 Contingency table

	−	+	
−	a	b	a + b
+	c	d	c + d
	a + c	b + d	

3.2 Variations of McNemar's Test

McNemar's test of symmetry is a large sample size testing procedure. When the number of observations for either b or c is small (say $b + c < 25$), the Chi-Square value is not well approximated by the Chi-Squared distribution. In such circumstances, the following two variations of McNemar's test are suggested.

3.2.1 Exact Binomial Test

The exact binomial test is an alternative to McNemar's test when the sample size is small. This basic idea of this test is to specify a binomial distribution instead of using Chi-Square distribution so that exact binomial test could be carried out to evaluate the imbalance in the discordant pairs of b (n_{12}) and c (n_{21}). There are two parameters for a binomial distribution, namely n and p. These two parameters are specified as $n = b + c$, and $p = 0.5$. To achieve a two-sided p-value, the p-value of the extreme tail is multiplied by 2. The formula to calculate exact p-value is as follows:

$$2 \sum_{i=0}^{b} \binom{n}{i} 0.5^i (1 - 0.5)^{n-i}$$

Edwards (1948) proposed the following continuity corrected version of the McNemar test to approximate the binomial exact p-value:

$$\chi^2 = \frac{(|b - c| - 1)^2}{b + c}$$

3.2.2 Mid-P McNemar's Test

The mid-p McNemar's test, also referred to as the mid-p binomial test, is another approach to address small sample size. The mid-p McNemar is the difference between the exact p-value and the pdf of binomial (n_{12}, n, 0.5). The formula for mid-p McNemar is stated below.

$\text{Exact } p - \text{value} = 2 \times \text{binomial cumulative density function} (n_{12}, n, 0.5)$

$\text{Mid} - \text{p value} = \text{Exact } p - \text{value} - \text{binomial probability density function} (n_{12}, n, 0.5)$

Table 3.2 below tabulates 112 Singapore citizens rating on the mayor's job performance. Each citizen rated the mayor as approve or disapprove in the first survey. Six months later, the second survey of the same citizens re-rates the mayor. The asymptotic Chi-Square is calculated as follows with *p*-value 0.033.

$$\chi^2 = \frac{(b-c)^2}{b+c} = \frac{(6-16)^2}{6+16} = 4.5455$$

The *p*-value of the continuity correction version of Chi-Square is 0.055.

$$\chi^2 = \frac{(|b-c|-1)^2}{b+c} = \frac{(|6-16|-1)^2}{6+16} = 3.6818$$

The function mcnemar.test() in the base, more specifically the stats package, performs the McNemar with correction by default. With the option specification, correct = FASLE, it gives the asymptotic McNemar test. The mcnemar.exact() function in package exact 2 × 2 produces the exact test. A function mcnemar.midp is written below to perform the mid-p test. Table 3.3 below summarizes the results of the four tests.

```
E1 <-
matrix(c(50, 16, 6, 40),
       nrow = 2,
       dimnames = list("1st Survey" = c("Approve", "Disapprove"),
                       "2nd Survey" = c("Approve", "Disapprove")))
E1
```

```
> E1
            2nd Survey
1st Survey  Approve Disapprove
  Approve        50          6
  Disapprove     16         40
```

```
# McNemar's Test with Correction
mcnemar.test(E1)
```

```
> mcnemar.test(E1)

        McNemar's Chi-squared test with continuity correction

data:  E1
McNemar's chi-squared = 3.6818, df = 1, p-value = 0.05501
```

Table 3.2 Survey on mayor

First survey	Second survey		
	Approve	Disapprove	Total
Approve	50	16	66
Disapprove	6	40	46
Total	56	56	112

Table 3.3 Summary of symmetry test

Test	Chi-Square	P-value
McNemar with continuity correction	3.68	0.055
McNemar without continuity correction	4.55	0.033
Exact binomial test	–	0.053
Mid-P McNemar test	–	0.035

```
# McNemar's Test without Correction
mcnemar.test(E1,correct=FALSE)
```

```
> mcnemar.test(E1,correct=FALSE)

        McNemar's Chi-squared test

data:  E1
McNemar's chi-squared = 4.5455, df = 1, p-value = 0.03301
```

```
# Exact Test
library(exact2x2)
mcnemar.exact(E1)
```

```
> mcnemar.exact(E1)

        Exact McNemar test (with central confidence intervals)

data:  E1
b = 6, c = 16, p-value = 0.05248
alternative hypothesis: true odds ratio is not equal to 1
95 percent confidence interval:
 0.1201837 1.0089245
sample estimates:
odds ratio
    0.375
```

```
# Mid-P Test - McNemar
mcnemar.midp <- function(b, c) {
    n <- b + c
    x <- min(b, c)
    p <- 2 * pbinom(x, n, .5, lower.tail=TRUE)
    midp <- p - dbinom(x, n, .5)
    return(midp)
}
```

```
mcnemar.midp(6,16)
```

```
> mcnemar.midp(6,16)
[1] 0.03468966
```

The exact binomial test gives p-value = 0.053 and McNemar's test with continuity correction gives Chi-Square = 3.68 and p-value = 0.055. The asymptotic McNemar's test gives Chi-Square = 4.55 and p-value = 0.033, and the mid-p McNemar's test gives p-value = 0.035. Both the McNemar's test and mid-p version provide stronger evidence for a statistically significant effect.

3.2.3   Which Test to Use?

The previous section illustrated four tests for symmetry. Which test should be the one to use? The general recommendation is to use the asymptotic test for large samples and the exact test for small samples. The argument for exact test is that the asymptotic test may violate the nominal significance level for small sample sizes because the required asymptotic property does not hold. The disadvantage of exact test is that it is conservative, producing unnecessary large p-values and with poor power (Fagerland et al 2013). The simulation results of Fagerland et al. (2013) suggest to use McNemar mid-p test and McNemar test without continuous correction as the latter is the most powerful test if the sample is not too small. They do not recommend the use of exact conditional test nor the asymptotic test with continuous correction.

3.3   Bower's Test of Symmetry

Bower's test of symmetry (1948) is a generalization of McNemar's test. Similar to McNemar's, Bowker's test is to access axial symmetry of a square a × a cross-tabulation. It is also Chi-Square asymptotic-based and hence is suitable for large sample. The hypothesis for Bower's is stated below in two different forms. The null hypothesis is that the probabilities of being classified into cells (i, j) and cell (j, i) are the same at the population level. Similar to McNemar's, Bowker's test of symmetry of a × a contingency table is also inherently 2-sided. The alternative hypothesis is undirected. The argument for marginal equality is the same for McNemar's test, i.e., if changes from one category to another are symmetrical, the marginal distributions stay the same.

$$H_0 : O_{ij} = O_{ji}$$
$$H_1 : O_{ij} \neq O_{ji} \text{ for at least one pair } O_{ij}, O_{ji}$$

or, alternatively

$$H_0 : \pi_{ij} = \pi_{ji}$$
$$H_1 : \pi_{ij} \neq \pi_{ji}$$

where $i \neq j$, $i = 1, 2, ..., k$, $j = 1, 2, ..., k$, and O_{ij} and O_{ji} are the frequencies of the symmetrical pairs in an $a \times a$ table.

The Q statistics of Bower's test of symmetry is computed as

$$Q = \sum_{i=1}^{k} \sum_{j=1}^{k} \frac{(n_{ij} - n_{ji})^2}{n_{ij} + n_{ji}} \quad \text{for } i > j$$

with $a(a - 1)/2$ degrees of freedom.

A survey was carried out asking 335 individuals whether they agreed with the policy that graduate mothers should be treated differently to give a subsidy for their newborn child. Within the same year, after the election, these same individuals were asked the same question. We would like to know if there is a change in the public opinion over the one-year period.

$$Q = \frac{(56 - 30)^2}{56 + 30} + \frac{(38 - 26)^2}{38 + 26} + \frac{(20 - 37)^2}{20 + 37} = 7.86 + 2.25 + 5.07 = 15.18$$
$$df = \binom{3}{2} = 3$$

The function mcnemar.test() not only applicable to 2×2 table for McNemar's test, it also runs the Bower's test of symmetry when an $a \times a$ tables with more than two rows and columns is specified. The following R syntax generates the Bower's test for Table 3.4.

```
opinion <- c("Agree","Disagree","Unsure")
Policy <- as.table(matrix(c(47, 56, 38,
                          30, 61, 20,
                          26, 37, 20), nrow = 3,
             dimnames = list(Before = opinion,
                             After = opinion)))
Policy
mcnemar.test(Policy)
```

Table 3.4 Before and after election

Before	After		
	Agree	Disagree	Unsure
Agree	47	30	26
Disagree	56	61	37
Unsure	38	20	20

```
> Policy
          After
Before     Agree Disagree Unsure
  Agree       47       30     26
  Disagree    56       61     37
  Unsure      38       20     20
> mcnemar.test(Policy)

        McNemar's Chi-squared test

data:  Policy
McNemar's chi-squared = 15.181, df = 3, p-value = 0.001669
```

Since the p-value is 0.0017, we reject the null hypothesis of symmetry. We expect at the population level, at least one pair of O_{ij} and O_{ji} differ. For instance, we do not expect the frequencies of those who disagree before and agree after is about the same as those disagree after and agree before.

3.4 Marginal Symmetry/Homogeneity Model

The marginal symmetry/homogeneity model (MS) is about the symmetry in the marginal total. When sum up the probabilities of cells in total by rows and columns, MS assumes the marginal totals are symmetric. Figure 3.1 shows the symmetry of the marginal totals by the double arrows line. For instance, π_{1+} represents the total probabilities of all the cells in row one, and π_{+1} represents the total probabilities of all the cells in column 1. MS hypothesizes that the total cells of first row have the same chance of the total cells of the first column; the total of cells of the second row has the same probability of the total cells of the second column and so on. If the MS model fits well, we would expect all the two marginal totals would be equal in probability for their respective rows and columns.

Formally, marginal homogeneity refers to equality between one or more of the row marginal proportions and the corresponding column proportion(s). The marginal symmetry/homogeneity model (MS) requires the row and column marginal distributions to be identical. The hypothesis of MS is as follows:

Fig. 3.1 Marginal total symmetry

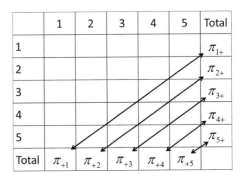

$$H_{MS} : \pi_{i+} = \pi_{+i} \text{ for all } i.$$

$$\text{where} \quad \pi_{i+} = \sum_j \pi_{ij}, \quad \pi_{+i} = \sum_j \pi_{ji}$$

Letting $d_i = \pi_{+i} - \pi_{i+}$ and $\mathbf{d} = (d_1, \ldots, d_{i-1})$, d_i becomes redundant in vector $\mathbf{d}$, hence generally there is i-1 degrees of freedom for testing marginal homogeneity (Sun and Yang 2008). Under the marginal homogeneity, the marginal homogeneity or the so-called Stuart-Maxwell test (Stuart 1955; Maxwell 1970), $E(\mathbf{d}) = 0$. For a 3×3 square table, there is a closed-form formula for the test statistics (Sun and Yang 2008; Walker 2002).

$$d_1 = (n_{12} + n_{13}) - (n_{21} + n_{31})$$
$$d_2 = (n_{21} + n_{23}) - (n_{12} + n_{32})$$
$$d_3 = (n_{31} + n_{32}) - (n_{13} + n_{23})$$
$$\bar{n}_{ij} = \frac{n_{ij} + n_{ji}}{2}, \quad \text{for } i \neq j$$
$$Z = \frac{\bar{n}_{23}d_1^2 + \bar{n}_{13}d_2^2 + \bar{n}_{12}d_3^2}{2(\bar{n}_{12}\bar{n}_{23} + \bar{n}_{12}\bar{n}_{13} + \bar{n}_{13}\bar{n}_{23})}$$

The following table tabulates the number of anomalies (0, 1, 2, 3) for 395 babies examined by Western and Chinese medical doctors. We are interested in testing whether the number of anomalies assessed by the Western medical doctors differs structurally from the number reported by the Chinese medical doctors (Table 3.5).

The mh_test() function from the base is used to carry out for marginal homogeneity model.

```
anomalies <- c(235, 23,  3, 0,
                21, 35,  8, 4,
                15, 11, 11, 1,
                 2,  1,  3, 8)
anomalies <- as.table(matrix(anomalies,
                ncol = 4, byrow=TRUE,
                dimnames = list(WMD = 0:3, CMD = 0:3)))
addmargins(anomalies)
mh_test(anomalies)
mh_test(anomalies, scores = list(response = c(0, 1, 2, 3)))
```

Table 3.5 Chinese and Western doctor diagnosis

Western medical doctor	Chinese medical doctor			
	0	1	2	3
0	235	23	3	0
1	21	35	8	4
2	15	11	11	1
3	2	1	3	8

```
> mh_test(anomalies)

        Asymptotic Marginal Homogeneity Test

data:   response by
        conditions (CMD, WMD)
        stratified by block
chi-squared = 5.1572, df = 3, p-value = 0.1606
```

The p-value of 0.16 indicates not rejecting the null hypothesis. However, the levels of the response are not treated as ordered. This can be included for analysis using the mh_test() function by the definition of an appropriate score as stated below. Similar result is found for specifying as ordered data.

```
> mh_test(anomalies, scores = list(response = c(0, 1, 2, 3)))

        Asymptotic Marginal Homogeneity Test for Ordered Data

data:   response (ordered) by
        conditions (CMD, WMD)
        stratified by block
Z = -2.0294, p-value = 0.04241
alternative hypothesis: two.sided
```

Alternatively, Stuart-Maxwell marginal homogeneity test could be carried out using StuartMAxwellTest() function from package DescTools.

Exercises

3.1 The following is a cross-tabulation of a project on interpretation and memorization of proverbs. There are 176 participants involved in this project. They have to interpret and memorize 20 proverbs. Ten of the proverbs are concretely worded, and the other 10 are abstractly worded. In the first part of the experiment, participants have to indicate whether they perceived a proverb as concrete or abstract in meaning. Participants are asked to rate proverbs as of one the categories: Concrete (=1), Intermediate (=2), or Abstract (=3) (von Eye et al. 1990; von Eye and Spiel 1996). The table below displays the 3×3 cross-tabulation of the proverb ratings. Carry out Bowker's test to examine whether there is a shift from interpreting proverbs as concrete is determined by proverb content (Table 3.6).

Table 3.6 Interpretation and memorization of proverbs

Proverb 1	Proverb 2		
	Concrete	Intermediate	Abstract
Concrete	11	2	8
Intermediate	4	5	12
Abstract	1	13	120

Table 3.7 Design matrix for axial symmetry

Cell	V1	V2	V3	V4	V5
(1,1)	1	0	0	0	0
(1,2)	0	0	1	0	0
(1,3)	0	0	0	1	0
(2,1)	0	0	1	0	0
(2,2)	0	1	0	0	0
(2,3)	0	0	0	0	1
(3,1)	0	0	0	1	0
(3,2)	0	0	0	0	1
(3,3)	0	0	0	0	0

3.2 Bower test can be equivalently expressed in terms of non-standard log-linear models. Use the following design matrix to run the table in 3.1 using non-standard log-linear model and compare the results of that in 3.1 (Table 3.7).

3.3 A program is designed to promote people to join public health profession. To evaluate the effectiveness of the program, 142 individuals were invited to participate in a study. Before the promotional program, 25 of them wish to join the public health profession. After the promotional program, 46 out of the 142 people wish to join the public health profession. Is there a significant change in the percentage of people who wish to join the public health profession? (Table 3.8).

3.4 About 88 patients characterized their craving of certain high-fat food products before and two weeks after an experimental diet therapy. The symptoms are coded as "never," "occasional," or "frequent." The research question is "Does the diet appears to have an effect on the frequency of these cravings?" The cell frequencies for symptom before and after treatment are tabulated below. Carry out marginal homogeneity test (Walker 2002; Sun and Yang 2008) (Table 3.9).

Table 3.8 Whether join public health profession

After	Before		
	Yes	No	Total
Yes	9	37	46
No	16	80	96
Total	25	117	142

Table 3.9 Symptoms before and after diet therapy

Pre-study	Two weeks			
	Never	Occasional	Frequent	Total
Never	14	6	4	24
Occasional	9	17	2	28
Frequent	6	12	8	26
Total	21	33	34	88

Table 3.10 Right and left eye vision

Right eye	Left eye				
	High grade	Second grade	Third grade	Lowest grade	Total
High grade	1520	266	124	66	1976
Second grade	234	1512	432	78	2256
Third grade	117	362	1772	205	2456
Lowest grade	36	82	179	492	789
Total	1907	2222	2507	841	7477

Calculate the value of Z using the following formula.

$$d_1 = (n_{12} + n_{13}) - (n_{21} + n_{31})$$
$$d_2 = (n_{21} + n_{23}) - (n_{12} + n_{32})$$
$$d_3 = (n_{31} + n_{32}) - (n_{13} + n_{23})$$
$$\bar{n}_{ij} = \frac{n_{ij} + n_{ji}}{2}, \quad \text{for } i \neq j$$
$$Z = \frac{\bar{n}_{23}d_1^2 + \bar{n}_{13}d_2^2 + \bar{n}_{12}d_3^2}{2(\bar{n}_{12}\bar{n}_{23} + \bar{n}_{12}\bar{n}_{13} + \bar{n}_{13}\bar{n}_{23})}$$

3.5 About 7477 women aged 30–39 tested with their right and left eyes on their unaided distance vision. The table below shows the results (Stuart 1953). Carry out marginal homogeneity test (Table 3.10).

References

Bower, A. H. (1948). A test for symmetry in contingency tables. *Journal of the American Statistical Association, 43,* 572–574.

Edward, A. L. (1948). Note on the "Correction for continuity" in testing the significance of the difference between correlated proportions. *Psychometrika, 13*(3), 185–187.

Fagerland, M. W., Lydersen, S., & Laake, P. (2013). The McNemar test for binary matched-pairs data: mid-p and asymptotic are better than exact conditional. *BMC Medical Research Methodology, 13,* 91.

Maxwell, A. E. (1970). Comparing the classification of subjects by two independent judges. *British Journal of Psychiatry, 116,* 651–655.

McMemar, Q. (1947). Note on the sampling error of the difference between correlated proportions or percentages, *Psychometrika, 12,* 153–157.

Stuart, A. (1953). The estimation and comparison of strengths of association in contingency tables. *Biometrika, 40*(1/2), 105–110.

Stuart, A. (1955). A test for homogeneity of the marginal distributions in a two-way classification, *Biometrika,* 412–416.

Sun, X. & Yang, Z. (2008). Generalized McNemar's test for homogeneity of the marginal distributions. SAS Global Forum 2008.

von Eye, A. & Spiel, C. (1996). Standard and non-standard log-linear symmetry models for measuring change in categorical variables. *American Statistician, 50,* 300–305.

von Eye, A., Jacobson, L. P., and Wills, S. D. (1990). *Proverbs: Inagery, Initerpretationi, and Memory*, 12th West Virginia University Conference on Life-Span Developmental Psychology.

Walker, G. A. (2002). *Common statistical methods for clinical research with SAS examples* (2nd ed.). SAS Institute Inc., Cary, North Carolina.

Chapter 4
Symmetry and Asymmetry Models

What exactly is a completely symmetry model? Table 4.1 shows an example of a perfectly and completely symmetric table. This table has the property of symmetric for all the off-diagonal cells in reference to the diagonal. For instance, the probability and the count of cell (1,2) are exactly the same count as cell (2,1) with both having a frequency of 2. The frequency of cell (2,3) and cell (3,2) also is having the same observation of 9. The rest of the cells in Table 4.1 exhibit the same property of count and probability symmetry. In general, an $a \times a$ contingency table with π_{ij} represents the probability that an observation falls into the (i,j) cell and the probability that an observation falls into the (j, i) cell, the chance of these two cells is equally likely, i.e., $\pi_{ij} = \pi_{ji}$ represents a perfectly symmetric table.

Completely symmetry model is important to look at because testing symmetry is often the preliminary step for examining a doubly classified table. This is one of the simplest models to explain about the association of cells. If the property of cell symmetry holds, we do not need to go further for more complicated model as this simple symmetry property explains the association of cells for a table. However, this model is rather restrictive, and in practice, the perfect symmetry model rarely comes by. Instead, departure from perfect symmetry is more often. This chapter covers the list of asymmetry models that are departed from this basic symmetry model. These set of models are referred to as asymmetry models because it exhibits variations from the symmetry.

4.1 Complete Symmetry Model

How do we know a table exhibits complete symmetry property? For a small table, it is easier to examine it with naked eye. For a large $a \times a$ doubly classified table, it is more difficult to examine for each and every cell. A 10×10 doubly classified table with 100 cells is difficult to examine with visual inspection. A formal procedure

© Springer Nature Singapore Pte Ltd. 2017
T.K. Tan, *Doubly Classified Model with R*,
https://doi.org/10.1007/978-981-10-6995-6_4

Table 4.1 Perfect symmetry table

	C1	C2	C3	Total
R1	5	2	8	15
R2	2	5	9	16
R3	8	9	5	22
Total	15	16	18	53

using nonstandard log-linear model specification to carry out doubly classified modeling is a better approach. A formal test, through hypothesis testing, assures us whether a specified doubly classified model fits the data with a qualified sampling error.

As mentioned in the last section, the complete symmetry model represents the baseline for a group of asymmetry models. Goodman (1985) referred it as the null-asymmetry model, specifically pointed out this model is the baseline for the asymmetry models. Complete symmetry model states that the probability of an observation falls in the cell (i,j) is equal to the probability falls in symmetric diagonal side of cell (j,i). For instance, the chance of an observation falls in cell $(1,2)$ is equally likely that it will fall in cell $(2,1)$, and the probability of an observation falls in cell $(2,4)$ is also equally likely it will fall in cell $(4,2)$ and so on. This implies that there are symmetry patterns of probabilities in cell position of a doubly classified table in reference to the principal diagonal for the entire body of a table. This principal diagonal refers to the diagonal cells that run from top left cell $(1,1)$ diagonally to the bottom right cell (a,a) for an $a \times a$ doubly classified table.

Now, we formally specify the model. For an $a \times a$ doubly classified table, π_{ij} represents the joint distribution of a doubly classified table for cells i and j, while π_{ji} represents the symmetric diagonal joint probability. The hypothesis of a complete symmetry model (S) is that the population probability of cell (i,j) is equal to cell (j, i) as shown below. The degrees of freedom for complete symmetry model is $a(a - 1)/2$.

$$H_S : \pi_{ij} = \pi_{ji} \quad \text{for } 1 \leq i \leq j \leq a$$

4.1.1 Symbolic Table

Completely symmetry model can also be represented in a table form with mathematical symbols within it, as shown in Fig. 4.1. This is referred to as a symbolic table. Throughout the text, this graphical symbolic table will be shown for all the doubly classified models. The main aim of a symbolic table is to aid readers to understand the basic characteristic and structure of a doubly classified model, on top of the usual mathematical representation.

Fig. 4.1 Symbolic table—
complete symmetry model

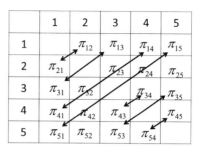

Figure 4.1 shows a 5 × 5 doubly classified table within each cell that indicates the probability of the cell (π_{ij}). For instance, cell (1,2) indicates the cell probability π_{12} and cell (4,5) with the cell probability π_{45}. The diagonal cells are intentionally left blank as they are of no relevancy to the complete symmetry model (S). These are cells that remain the same value under complete symmetry modeling; i.e., the estimated values of these cells are the same as the raw data original value. The seven double arrows show the symmetry equality of the probability of the cells. For instance, $\pi_{12} = \pi_{21}$, $\pi_{13} = \pi_{31}$, and $\pi_{14} = \pi_{41}$. Only seven double arrows instead of all the 10 cells relationship with 10 arrows are indicated in the symbolic table. This is to reduce confusion in vision of too many arrows.

Symbolic tables are illustrated throughout the whole text, one symbolic table for one doubly classified model. As not all readers are mathematically inclined, the symbolic tables help to better understand by combining mathematical symbols with table to create a new form of representation to aid readers to grasp the association of cells for doubly classified model. In a nutshell, as a doubly classified model describes the relationship of cells in term of probabilities, odds, odds ratios, log odds ratios of a table, using mathematic symbols in combination of table form, for easy explanation, gives the symbolic table. The lists of mathematic symbols that are used in a symbolic table are listed below. Section 10.1 lists all the symbolic tables used in this book.

Symbol	Description
π_{ij}	Probability for cell (i,j)
π_{i+}	Row total
π_{+j}	Column total
θ_{ij}	Odds ratio
Δ_{lower}	Expected group cells of lower diagonal
Δ_{upper}	Expected group cells of upper diagonal
δ_k	$\delta_k = \pi_{ij}/\pi_{ji}$ where $k = j - i$
Φ_{ij}	Log of odds ratio $\ln(\theta_{ij})$
$\leftrightarrow$	Equality

4.1.2   Nonstandard Log-Linear Approach in Generating Model

How to model a complete symmetry model or more generally, a doubly classified model? Lawal (2000, 2001, 2003, 2004) and von Eye and Spiel (1996) advocate using nonstandard log-linear model to generate doubly classified models. The basic idea of this approach is to employ the generalized linear model (GLM) with the correct specification of factor(s) and variable(s) to generate the required doubly classified model. In fact, to fit an $a \times a$ doubly classified model, any software package that contains GLM function will be suffice to generate a doubly classified table model by specifying the right factor(s) and variable(s). Lawal used SAS PROC GENMOD and SPSS GENLOG to fit the various doubly classified models. This book concentrates on package R function glm() to fit doubly classified table.

For an $a \times a$ square table, a nonstandard log-linear model written in the generalized linear form is as follows (Cogg et al. 1990; von Eye and Spel 1996; Lawal 2001).

$$\ell_{ij} = \ln\left(m_{ij}\right)$$
$$\ell(\mathbf{m}) = \mathbf{X}\lambda$$

where $\mathbf{X}$ is a design matrix consisting of 0 s and 1 s, and λ is a vector of parameters, and $\mathbf{m}$ is a a^2 vector of expected values under some specified doubly classified model. The underlying distribution of the random component is the Poisson distribution with a log link. The general R functions to generate doubly classified model are glm() (generalized linear model) and gnm() (generalized nonlinear model) with the following syntax,

$$\text{glm}(m \sim f1 + f2 + \cdots + fn, \text{family} = \text{poisson}$$
$$\text{gnm}(m \sim f1 + f2 + \cdots + fn, \text{family} = \text{poisson}$$

where $f1$, $f2$ to fn are factors or variables. glm() is the basic function used for the whole book although gnm() function also serves the same purpose with similar syntax specification.

4.1.3   Complete Symmetry Model—Nonstandard Log-Linear Specification

The nonstandard log-linear formulation of complete symmetry model (S) is in the following form.

Table 4.2 Specifying factor values for a complete symmetry model

S	Condition	Value
1	$i = j = 1$	1
2	$(i,j) = (1,2) = (2,1)$	2
3	$(i,j) = (1,3) = (3,1)$	3
4	$(i,j) = (1,4) = (4,1)$	4
5	$i = j = 2$	5
6	$(i,j) = (2,3) = (3,2)$	6
7	$(i,j) = (2,4) = (4,2)$	7
8	$i = j = 3$	8
9	$(i,j) = (3,4) = (4,3)$	9
10	$i = j = 4$	10

$$\ln\left(m_{ij}\right) = \mu + \lambda_{ij}^S$$

The above assignment gives the S matrix below.

$$S = \begin{bmatrix} 1 & 2 & 3 & 4 \\ 2 & 5 & 6 & 7 \\ 3 & 6 & 8 & 9 \\ 4 & 7 & 9 & 10 \end{bmatrix}$$

The assignment of values to the above S matrix is illustrated in Table 4.2. First, we have to specify factors that describe the relationship of the cells that stipulates the complete symmetry model. The factor takes on a unique value for each diagonal cell and a unique value of each pair of cells that are diagonally symmetrical. For a 4 × 4 cross-tabulation, since we are not concerned about the diagonal cells, each of the cells (1,1), (2,2), (3,3), and (4,4) is assigned with a unique values 1, 5, 8, and 10, respectively. The symmetry cell (1,2) and symmetry cell (2,1) are assigned with a common value of 2. Similarly, the symmetrical cell (1,3) and symmetrical cell (3,1) are assigned with a common value of 3. This process goes on for the rest of the cells that are in symmetrical position in reference to the diagonal of a table.

The S factor could be generated using the below formula for $a \times a$ table (Lawal and Sundheim 2002; Lawal 2004).

$$S_{ij} = \begin{cases} (k+1) - (i+1)(\frac{1}{2}i+1) + (a+3)(i+1) - 3 - 2a & i<j \\ (k+1) - (j+1)(\frac{1}{2}j+1) + (a+3)(j+1) - 3 - 2a & i\geq j \end{cases}$$

4.1.4 Properties of Complete Symmetry Model

Given the probability is in symmetry for the complete symmetry model, the sum over rows has the same probability of sum over columns. That is, the probability summing over rows π_{i+} is the same probability that it sums over columns π_{+i}. We refer this as marginal homogeneity, a basic characteristics of complete symmetry model.

$$\pi_{i+} = \sum_j \pi_{ij} = \sum_j \pi_{ji} = \pi_{+i}$$

The expected frequency of complete symmetry model is $\ln \mu_{ij} = n\pi_{ij}$. The MLE estimates of the frequency are the average of the two symmetric cells.

$$\ln \mu_{ij} = \lambda + \lambda_i + \lambda_j + \lambda_{ij}$$

$$\text{where} \quad \lambda_{ij} = \lambda_{ji}$$

$$\hat{\mu}_{ij} = \frac{n_{ij} + n_{ji}}{2}$$

4.1.5 Function Model.Summary

The following function called model.summary() is used throughout the text to output the results of doubly classified model. It calculates fit statistics AIC, BIC, G^2, degrees of freedom, p-value, and regression output. It also outputs raw data, estimated values, residuals, odds ratios, and log odds ratios.

The model.summary reads in a lm object (obj) and produces fit statistics G^2, AIC, BIC, p-values which are stored into a matric called result. Five matrices are created to output the raw table, estimated values, residuals, local odds ratios, and log local odds ratios. The names of these five matrices are Table, PredictedTAble, ResidualTable, LocalOddsRatios, and LogOddsRatios, respectively.

```
model.summary <- function(obj) {
    # Calculate AIC, BIC, and P Value
    G2 <- round(obj$deviance,2)
    aic <- round(obj$deviance-obj$df.residual*2,2)
    bic <- round(obj$deviance-obj$df.residual*log(sum(obj$y)),2)
    p <- round(pchisq(obj$deviance, obj$df.residual, lower.tail=F),4)
    result<-matrix(0,1,5)
    rownames(result)<-""
    colnames(result)<c("Df","G2","p","AIC","BIC")
    result[1,]<-c(obj$df.residual,G2,p,aic,bic)
    # Raw Data
    y <- obj$y
    # Fitted Values and Residuals
    fitted.values <- obj$fitted.values
```

```
residuals <- obj$residuals

fit.res.y <- rbind(y,fitted.values,residuals)
 # Dimension of table
a <- sqrt(length(obj$y))

# Put Raw Table as a matirx
Table <- matrix(y,nrow=a,ncol=a,byrow=T)

# Put Estimated Table as a matirx
PredictedTable <- matrix(fitted.values,nrow=a,ncol=a,byrow=T)

# Put Estimated Residual as a Table
ResidualTable <- matrix(residuals,nrow=a,ncol=a,byrow=T)

# Calculate Odds Ratios and Local Odds Ratios

Fitted <- matrix(fitted(obj),nrow=a,ncol=a,byrow=T)
LocalOddsRatios <- matrix(data=NA,nrow=a-1,ncol=a-1)
for (i in 1:a-1) {
    for (j in 1:a-1) {
        LocalOddsRatios[i,j] <- (Fitted[i,j] * Fitted[i+1,j+1]) /
                                (Fitted[i+1,j] * Fitted[i,j+1])
    }
}
LogOddsRatios <- log(LocalOddsRatios)

print("Raw Data, Fitted Values and Residuals:")
print(round(fit.res.y,2))
print("Raw Table")
print(Table)
print("Estimated Table")
print(PredictedTable)
print("Residual Table")
print(ResidualTable)
print("Local Odds Ratios Table")
print(LocalOddsRatios)
print("Log Local Odds Ratios Table")
print(LogOddsRatios)
print("Model Summary:")
return(result)
}
```

Table 4.3 Residence mobility pattern

Last residence	Current residence			
	Ang Mo Kio	Toa Payoh	Yishun	Sembawang
Ang Mo Kio	150	66	24	33
Toa Payoh	68	12	32	81
Yishun	20	32	72	20
Sembawang	38	82	12	42

The following is a 4 × 4 doubly classified table that records the movement of residents for the four Singapore towns, namely Ang Mo Kio, Toa Payoh, Yishun, and Sembawang. Their movements from last residence to their current residence are captured in Table 4.3.

The *R* program to run the complete symmetry model for the above table is shown below. ComS is a vector with 16 elements that contains the data values. *s* is a vector that specifies the *s* matrix discussed above for the parameter λ_{ij}^{S}. It is subsequently converted into a factor using the as.factor() function. The glm() function generates the GLM, specified as Poisson family. The dependent variable is ComS, and the independent is the s factor. The results of the model are stored in S. The command summary(S) prints the results of the model. The command model.summary(S) invokes the function model.summary() and prints the model fit and the five matrices mentioned produced by model.summary() function for the complete symmetry model.

```
ComS <- c(150, 66, 24, 33,
           68, 12, 32, 81,
           20, 32, 72, 20,
           38, 82, 12, 42)
s<-c( 1, 2, 3, 4,
      2, 5, 6, 7,
      3, 6, 8, 9,
      4, 7, 9,10)
s<-as.factor(s)
S<-glm(ComS~s,family=poisson)
summary(S)
model.summary(S)
```

```
> summary(S)

Call:
gnm(formula = ComS ~ s, family = poisson)

Deviance Residuals:
      Min        1Q    Median        3Q       Max
  -1.04672  -0.07220   0.00000   0.07196   0.96215

Coefficients:
             Estimate Std. Error z value Pr(>|z|)
(Intercept)   5.01064    0.08165  61.367  < 2e-16 ***
s2           -0.80594    0.11887  -6.780 1.20e-11 ***
s3           -1.91959    0.17145 -11.196  < 2e-16 ***
s4           -1.44110    0.14405 -10.004  < 2e-16 ***
s5           -2.52573    0.30000  -8.419  < 2e-16 ***
s6           -1.54490    0.14930 -10.347  < 2e-16 ***
s7           -0.61003    0.11314  -5.392 6.98e-08 ***
s8           -0.73397    0.14337  -5.119 3.07e-07 ***
s9           -2.23805    0.19472 -11.494  < 2e-16 ***
s10          -1.27297    0.17457  -7.292 3.06e-13 ***
---
Signif. codes:  0 '***' 0.001 '**' 0.01 '*' 0.05 '.' 0.1 ' ' 1

(Dispersion parameter for poisson family taken to be 1)

Residual deviance: 2.7739 on 6 degrees of freedom
AIC: 110.71
```

The above output prints the estimates of the complete symmetry model. The value of μ (intercept) represents the estimated value of cell (1,1), $\exp(5.01064) = 150$, which is the same as the original data of cell (1,1). As s is a factor, $s1$ does not appear in the output as this is the reference dummy. Thus, the intercept represents $s1$, the estimated value for cell (1,1). Intercept represents the non-movers of those lived in Ang Mio Kio. Similarly, the fitted values of $s5$, $s8$, $s10$ have the same as those of their original raw values. For instance, $s10 : \exp(5.01064 - 1.27297) = 42$ which is the same original table value. The estimated values of the rest of the factors ($s2$–$s10$) show the equality of the cell symmetry. For instance, $s2$ is the estimated value of cell (1,2) and cell (2,1), $s2 : \exp(5.01064 - 0.80594) = 67$, and $s3$ is the estimated value of cell (1,3) and cell (3,1), $s3 : \exp(5.01064 - 1.91959) = 22$. The rest of the estimated values can be interpreted accordingly.

The first output printed under the model.summary() contains the raw table data (y), the fitted values (fitted.values), and the residuals (residuals). We can observe that the raw data are not very different from the fitted values. For instance, the second element of the y (66) is just one value away from the fitted value of 67. Almost all the residuals are small with the exception of elements 12 and 15.

```
[1] "Raw Data, Fitted Values and Residuals:"
                 1     2     3     4    5  6  7     8     9 10 11    12    13    14    15 16
y              150 66.00 24.00 33.00 68.00 12 32 81.00 20.00 32 72 20.00 38.00 82.00 12.00 42
fitted.values  150 67.00 22.00 35.50 67.00 12 32 81.50 22.00 32 72 16.00 35.50 81.50 16.00 42
residuals        0 -0.01  0.09 -0.07  0.01  0  0 -0.01 -0.09  0  0  0.25  0.07  0.01 -0.25  0
```

Next, five matrices are printed. The "Raw Table" reproduces the original table data. The "Estimated Table" produces the estimated values. The "Residual Table" prints the differences between raw and estimated values. The "Local Odds Ratios Table" produces the estimated odds ratios, and the "Log Local Odds Ratios Table" prints the log odds ratios.

```
[1] "Raw Table"
      [,1] [,2] [,3] [,4]
[1,]   150   66   24    33
[2,]    68   12   32    81
[3,]    20   32   72    20
[4,]    38   82   12    42
[1] "Estimated Table"
       [,1] [,2] [,3] [,4]
[1,] 150.0 67.0   22 35.5
[2,]  67.0 12.0   32 81.5
[3,]  22.0 32.0   72 16.0
[4,]  35.5 81.5   16 42.0
[1] "Residual Table"
                [,1]            [,2]             [,3]             [,4]
[1,] -5.684342e-16 -1.492537e-02  9.090909e-02 -7.042254e-02
[2,]  1.492537e-02  4.440892e-16  0.000000e+00 -6.134969e-03
[3,] -9.090909e-02  0.000000e+00  7.894919e-16  2.500000e-01
[4,]  7.042254e-02  6.134969e-03 -2.500000e-01  1.184238e-15
[1] "Local Odds Ratios Table"
           [,1]       [,2]        [,3]
[1,] 0.4009802 8.1212121  1.5783451
[2,] 8.1212121 0.8437500  0.0872529
[3,] 1.5783451 0.0872529 11.8125000
[1] "Log Local Odds Ratios Table"
            [,1]        [,2]        [,3]
[1,] -0.9138433  2.094479  0.4563769
[2,]  2.0944794 -0.169899 -2.4389445
[3,]  0.4563769 -2.438945  2.4691583
```

The fit statistics is the last output from model.summary(). This section prints the fit statistics with deviance $G^2 = 2.77$, degrees of freedom = 6, p-value = 0.8366, AIC = -9.23, and BIC = -37.21. Since the p-value is greater than 0.05, we do not reject the hypothesis that the complete symmetry model fits well.

```
[1] "Model Summary:"
 Df   G2       p    AIC     BIC
  6 2.77 0.8366 -9.23 -37.21
```

The estimated values, $\hat{m}_{ij}$, and the estimated odds ratios, $\hat{\theta}_{ij}$, of complete symmetry model are reproduced below.

$$\hat{m}_{ij} = \begin{bmatrix} 150 & 67 & 22 & 35.5 \\ 67 & 12 & 32 & 81.5 \\ 22 & 32 & 72 & 16 \\ 35.5 & 81.5 & 16 & 42 \end{bmatrix}$$

$$\hat{\theta}_{ij} = \begin{bmatrix} 0.40 & 8.12 & 1.58 \\ 8.12 & 0.84 & 0.09 \\ 1.58 & 0.09 & 11.81 \end{bmatrix}$$

$$\frac{\hat{\theta}_{12}}{\hat{\theta}_{21}} = \frac{8.12}{8.12} = 1$$

The estimated values show the symmetrical pattern of complete symmetry model. The value of $\hat{m}_{ji}$ is the same as the value of $\hat{m}_{ij}$ as stipulated by complete symmetry model. For instance, $\hat{m}_{12}$ cell (1,2) and $\hat{m}_{21}$ cell (2,1) have the same estimated value of 67, and $\hat{m}_{23}$ cell (2,3) and $\hat{m}_{32}$ cell (3,2) have the same value of

32. Since the estimated values are in symmetry, the odds ratios are also in symmetry. For instance, $\hat{\theta}_{12}$ is the same as $\hat{\theta}_{21}$ with value of 8.12. Hence, the ratio of $\hat{\theta}_{12}/\hat{\theta}_{21}$ is equal to 1. It is noted that the MLE estimates of $\hat{m}_{ij}$ are $n_{ij}n_{ji}/2$. For instance, the estimated value of $\hat{m}_{12} = \hat{m}_{21} = (n_{12} + n_{21})/2 = (68 + 66)/2 = 67$.

4.1.6   A Detailed Illustration—The Dummy Specification

This section gives a detailed explanation on how doubly classified model is carried out under R, the dummy specification. For a 4×4 doubly classified table, to model complete symmetry model, it requires 10 dummy variables Z_{1ij} to Z_{10ij} with λ_{10}^S sets to zero, resulting in 9 dummy variables. It is not necessary to set λ_{10}^S as zero. Any of the dummy variables can be set to zero. The default of R is that it sets the first factor as zero. The specification of the dummy variables is according to the s matrix numbering. Those with the same number are grouped as a dummy. For instance, cell (1,2) and cell (2,1) are specified under Z_{2ij}, and cell (2,3) and cell (3,2) are specified under Z_{6ij}.

$$l_{ij} = \mu + \lambda_1^S Z_{1ij} + \lambda_2^S Z_{2ij} + \lambda_3^S Z_{3ij} + \lambda_4^S Z_{4ij} + \lambda_5^S Z_{5ij} + \lambda_6^S Z_{6ij}$$
$$+ \lambda_7^S Z_{7ij} + \lambda_8^S Z_{8ij} + \lambda_9^S Z_{9ij} + \lambda_{10}^S Z_{10ij}$$

$$Z_{1ij} = \begin{cases} 1 & \text{cell(1,1)} \\ 0 & \text{otherwise} \end{cases} \qquad Z_{6ij} = \begin{cases} 1 & \text{cell(2,3),cell(3,2)} \\ 0 & \text{otherwise} \end{cases}$$

$$Z_{2ij} = \begin{cases} 1 & \text{cell(1,2),cell(2,1)} \\ 0 & \text{otherwise} \end{cases} \qquad Z_{7ij} = \begin{cases} 1 & \text{cell(2,4),cell(4,2)} \\ 0 & \text{otherwise} \end{cases}$$

$$Z_{3ij} = \begin{cases} 1 & \text{cell(1,3),cell(3,1)} \\ 0 & \text{otherwise} \end{cases} \qquad Z_{8ij} = \begin{cases} 1 & \text{cell(3,3)} \\ 0 & \text{otherwise} \end{cases}$$

$$Z_{4ij} = \begin{cases} 1 & \text{cell(1,4),cell(4,1)} \\ 0 & \text{otherwise} \end{cases} \qquad Z_{9ij} = \begin{cases} 1 & \text{cell(3,4),cell(4,3)} \\ 0 & \text{otherwise} \end{cases}$$

$$Z_{5ij} = \begin{cases} 1 & \text{cell(2,2)} \\ 0 & \text{otherwise} \end{cases} \qquad Z_{10ij} = \begin{cases} 1 & \text{cell(4,4)} \\ 0 & \text{otherwise} \end{cases}$$

$$\lambda_1^S = 0$$

Because of the structure of s, there are 11 parameters including the common intercept. The nonstandard log-linear model is thus has the following matrix structure.

$$
\begin{bmatrix}
l_{11} \\
l_{12} \\
l_{13} \\
l_{14} \\
l_{21} \\
l_{22} \\
l_{23} \\
l_{24} \\
l_{31} \\
l_{32} \\
l_{33} \\
l_{34} \\
l_{41} \\
l_{42} \\
l_{43} \\
l_{44}
\end{bmatrix}
=
\begin{bmatrix}
1 & 1 & 0 & 0 & 0 & 0 & 0 & 0 & 0 & 0 & 0 \\
1 & 0 & 1 & 0 & 0 & 0 & 0 & 0 & 0 & 0 & 0 \\
1 & 0 & 0 & 1 & 0 & 0 & 0 & 0 & 0 & 0 & 0 \\
1 & 0 & 0 & 0 & 1 & 0 & 0 & 0 & 0 & 0 & 0 \\
1 & 0 & 1 & 0 & 0 & 0 & 0 & 0 & 0 & 0 & 0 \\
1 & 0 & 0 & 0 & 0 & 1 & 0 & 0 & 0 & 0 & 0 \\
1 & 0 & 0 & 0 & 0 & 0 & 1 & 0 & 0 & 0 & 0 \\
1 & 0 & 0 & 0 & 0 & 0 & 0 & 1 & 0 & 0 & 0 \\
1 & 0 & 0 & 1 & 0 & 0 & 0 & 0 & 0 & 0 & 0 \\
1 & 0 & 0 & 0 & 0 & 0 & 1 & 0 & 0 & 0 & 0 \\
1 & 0 & 0 & 0 & 0 & 0 & 0 & 0 & 1 & 0 & 0 \\
1 & 0 & 0 & 0 & 0 & 0 & 0 & 0 & 0 & 1 & 0 \\
1 & 0 & 0 & 0 & 1 & 0 & 0 & 0 & 0 & 0 & 0 \\
1 & 0 & 0 & 0 & 0 & 0 & 0 & 1 & 0 & 0 & 0 \\
1 & 0 & 0 & 0 & 0 & 0 & 0 & 0 & 0 & 1 & 0 \\
1 & 0 & 0 & 0 & 0 & 0 & 0 & 0 & 0 & 0 & 1
\end{bmatrix}
\begin{bmatrix}
\mu \\
\lambda_1 \\
\lambda_2 \\
\lambda_3 \\
\lambda_4 \\
\lambda_5 \\
\lambda_6 \\
\lambda_7 \\
\lambda_8 \\
\lambda_9 \\
\lambda_{10}
\end{bmatrix}
$$

The above formulation is the familiar regression model with the design matrix consisting of 0s and 1s. It is equivalent to the above s factor specification. For instance, the third column of 1s appears on cell (1,2) and cell (2,1) that are corresponding position of the s factor.

The R syntax to generate the complete symmetry model is stated below. The set of 10 vectors S11 to S34_43 specify the dummy variables for s factor. There are two models specified below. The first model sets S11 λ_1 as zero that generates the output same as the previous output specified earlier. The second model sets S44 λ_{10} as zero. Although the estimated coefficients are different from the first model, both models produce the same results and same fit statistics.

```
S11     <- c(1,0,0,0,0,0,0,0,0,0,0,0,0,0,0,0)
S22     <- c(0,0,0,0,0,1,0,0,0,0,0,0,0,0,0,0)
S33     <- c(0,0,0,0,0,0,0,0,0,0,1,0,0,0,0,0)
S44     <- c(0,0,0,0,0,0,0,0,0,0,0,0,0,0,0,1)
S12_21 <- c(0,1,0,0,1,0,0,0,0,0,0,0,0,0,0,0)
S13_31 <- c(0,0,1,0,0,0,0,0,1,0,0,0,0,0,0,0)
S14_41 <- c(0,0,0,1,0,0,0,0,0,0,0,0,1,0,0,0)
S23_32 <- c(0,0,0,0,0,0,1,0,0,1,0,0,0,0,0,0)
S24_42 <- c(0,0,0,0,0,0,0,1,0,0,0,0,0,1,0,0)
S34_43 <- c(0,0,0,0,0,0,0,0,0,0,0,1,0,0,1,0)


# Set S11=0
CompleteSys <- glm(ComS ~ S22 + S33 + S44 +
                         S12_21 + S13_31 + S14_41 +
                         S23_32 + S24_42 + S34_43, family=poisson)
```

```
summary(CompleteSys)
model.summary(CompleteSys)

# Set S44=0
CompleteSys <- glm(ComS ~ S11 + S22 +160;S33 +
                        S12_21 + S13_31 + S14_41 +
                        S23_32 + S24_42 + S34_43, family=poisson)
summary(CompleteSys)
model.summary(CompleteSys)
```

The outputs for the first model are printed below, which are exactly the same as the previous model output using s as a factor.

```
Coefficients:
              Estimate Std. Error z value Pr(>|z|)
(Intercept)   5.01064    0.08165  61.367  < 2e-16 ***
S22          -2.52573    0.30000  -8.419  < 2e-16 ***
S33          -0.73397    0.14337  -5.119 3.07e-07 ***
S44          -1.27297    0.17457  -7.292 3.06e-13 ***
S12_21       -0.80594    0.11887  -6.780 1.20e-11 ***
S13_31       -1.91959    0.17145 -11.196  < 2e-16 ***
S14_41       -1.44110    0.14405 -10.004  < 2e-16 ***
S23_32       -1.54490    0.14930 -10.347  < 2e-16 ***
S24_42       -0.61003    0.11314  -5.392 6.98e-08 ***
S34_43       -2.23805    0.19472 -11.494  < 2e-16 ***
```

The following prints the second model. As the second model sets S44 as zero, the estimated coefficients differ from the first model. The value of μ (intercept) represents the estimated value of cell (4,4), $\exp(3.7377) = 42$, which is the same as the original data of cell (4,4). Similarly, the fitted values of $s1$ cell (1,1) are the same as their original values, $s11 : \exp(3.7377 + 1.2730) = 150$.

```
Coefficients:
              Estimate Std. Error z value Pr(>|z|)
(Intercept)   3.7377     0.1543   24.223  < 2e-16 ***
S11           1.2730     0.1746    7.292 3.06e-13 ***
S22          -1.2528     0.3273   -3.827 0.000130 ***
S33           0.5390     0.1942    2.776 0.005503 **
S12_21        0.4670     0.1768    2.641 0.008268 **
S13_31       -0.6466     0.2157   -2.997 0.002722 **
S14_41       -0.1681     0.1947   -0.864 0.387737
S23_32       -0.2719     0.1986   -1.369 0.170880
S24_42        0.6629     0.1730    3.831 0.000128 ***
S34_43       -0.9651     0.2346   -4.113 3.91e-05 ***
```

4.2 Conditional Symmetry Model

When the complete symmetry model does not hold, we look for other doubly classified models. Conditional symmetry model is the closest to the complete symmetry model, first introduced by Caussinus (1965). The main characteristic of this model is that the probability of the upper triangular cells is a constant to that of the probability of the lower triangular cells. The hypothesis of the conditional symmetry model is thus specified as follows:

$$H_{CS} : \pi_{ij} = \gamma \pi_{ji} \quad \text{for} \ i < j$$

The following symbolic table restates the above hypothesis. The double arrow lines show the relationships of the upper triangular cell and lower triangular cell that are in a constant term γ. For instance, $\pi_{12} = \gamma \pi_{21}$, and $\pi_{13} = \gamma \pi_{31}$. This property applies for the rest of the symmetrical cells (Fig. 4.2).

Another way of looking at conditional symmetry model is to express symmetry in total form for the off-diagonal cells. By representing the sum of upper diagonal cells as Δ_{upper} and the sum of the lower diagonal as Δ_{lower}, the conditional symmetry (CS) model specifies the equality of the two expected group cells ($\Delta_{upper} = \gamma \Delta_{lower}$) by a constant term γ. This property is shown in Fig. 4.3. Conditional symmetry model preserves both the ratio γ of triangle totals and the individual cell.

There are a number of ways to specify the conditional symmetry model. McCullagh (1978) defines conditional symmetry as follows:

$$H_{CS} : \pi_{ij} = \begin{cases} \gamma \phi_{ij} & i < j \\ \phi_{ij} & i \geq j \end{cases}$$

where $\phi_{ij} = \phi_{ji}$ and π_{ij} denote the probability that an observation falls in ith row and jth column of the table. When $\gamma = 1$, conditional symmetry model becomes

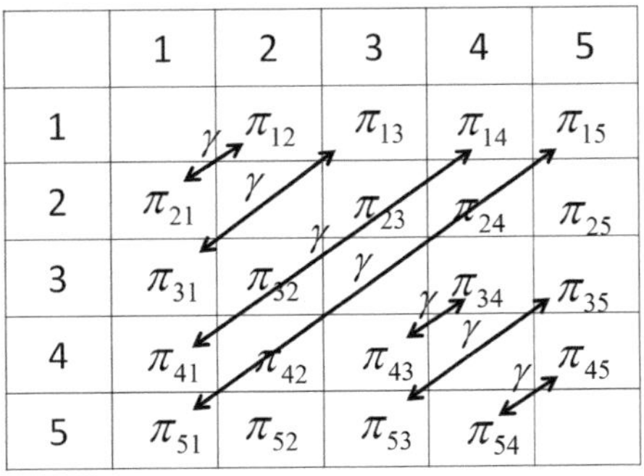

Fig. 4.2 Conditional symmetry model

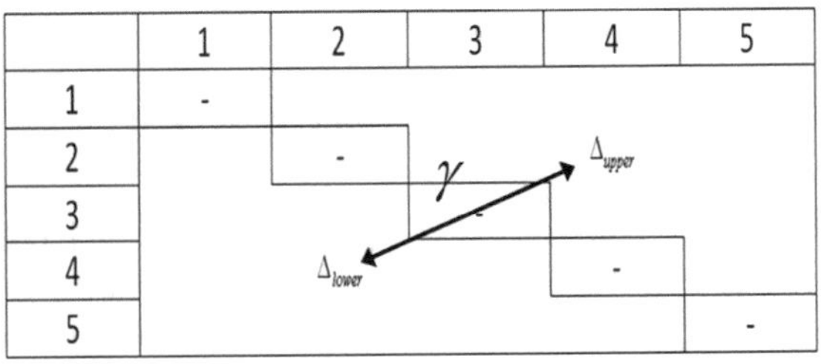

Fig. 4.3 Conditional symmetry model: Δ_{upper} and Δ_{lower}

complete symmetry model. Alternatively, the conditional symmetry model can also
be expressed as follows:

$$H_{CS} : \frac{\pi_{ij}}{\pi_{ji}} = \gamma \quad i<j$$

This way of formulation states that the ratio of the probability that an observation
falls in cell (i,j) in the upper right triangle of the table to the probability that the
observation falls in cell (j,i) in the lower left triangle is a constant term γ for $i <
j$. Yet another specification of the conditional symmetry model can be expressed as
follows:

$$H_{CS} : \pi_{ij}^{U} = \pi_{ji}^{L} \quad i<j$$

where

$$\pi_{ij}^{U} = \frac{\pi_{ij}}{\delta_U}, \quad \pi_{ji}^{L} = \frac{\pi_{ji}}{\delta_L}, \quad \delta_U = \sum_{s<t}\sum \pi_{st} \; \delta_L = \sum_{s>t}\sum \pi_{st}.$$

The above specifications indicate that the probability that an observation falls in
cell (i,j) on condition that it falls in one of the cells in the upper right triangle of the
table is equal to the probability that the observation falls in cell (j,i) on condition
that it falls in one of the cells in the lower left triangle of the table. The conditional
symmetry model thus shows a structure of symmetry between two conditional
distributions π_{ij}^{U} and π_{ji}^{L}.

The nonstandard log-linear model specification of conditional symmetry model
with $(a+1)(a-1)/2$ degrees of freedom is as follows:

$$\ln\left(\hat{m}_{ij}\right) = \mu + \lambda_k^{S_{ij}} + \eta^{CS} \quad \text{for } i \neq j$$

$$\Delta_{\text{upper}} = \sum_{i<j} f_{ij}, \quad \Delta_{\text{lower}} = \sum_{i>j} f_{ij}$$

$$CS_{ij} = \begin{cases} 2 & i>j \\ 3 & i<j \\ 1 & i=j \end{cases} \quad \text{or} \quad CS_{ij} = \begin{cases} 1 & i \le j \\ 2 & i>j \end{cases}$$

$$S = \begin{bmatrix} 1 & 2 & 3 & 4 \\ 2 & 5 & 6 & 7 \\ 3 & 6 & 8 & 9 \\ 4 & 7 & 9 & 10 \end{bmatrix}, \quad CS = \begin{bmatrix} 1 & 3 & 3 & 3 \\ 2 & 1 & 3 & 3 \\ 2 & 2 & 1 & 3 \\ 2 & 2 & 2 & 1 \end{bmatrix} \quad \text{or} \quad CS = \begin{bmatrix} 1 & 1 & 1 & 1 \\ 2 & 1 & 1 & 1 \\ 2 & 2 & 1 & 1 \\ 2 & 2 & 2 & 1 \end{bmatrix}$$

The following 4×4 doubly classified table records the right and left eye distant
vision of 635 patients from an eye clinic. The vision of eye is classified into four
grades. The higher the grade, the better is the eye sight (Table 4.4).

78

Symmetry and Asymmetry Models

Right eye grade	Left eye grade			
	1	2	3	4
1	50	21	24	35
2	42	22	12	42
3	56	32	52	16
4	67	82	20	62

Table 4.4 Right and left eye distant vision

The following R syntax generates the complete symmetry and conditional symmetry models of the above table. The difference between complete symmetry and conditional symmetry is the inclusion of the factor CS in the glm() function.

```
ConS <- c(50, 21, 24, 35,
          42, 22, 12, 42,
          56, 32, 52, 16,
          67, 82, 20, 62)
s<-c( 1, 2, 3, 4,
      2, 5, 6, 7,
      3, 6, 8, 9,
      4, 7, 9,10)
s<-as.factor(s)
CS <- c( 1, 1, 1, 1,
         2, 1, 1, 1,
         2, 2, 1, 1,
         2, 2, 2, 1)
CS <- as.factor(CS)
CSSys<-glm(ConS~s+CS,family=poisson)
summary(CSSys)
model.summary(CSSys)
```

The estimated coefficients of conditional symmetry model are printed below.

```
Coefficients:
            Estimate Std. Error z value Pr(>|z|)
(Intercept)  3.91202    0.14142  27.662  < 2e-16 ***
s2          -0.86528    0.20078  -4.310 1.64e-05 ***
s3          -0.62638    0.19220  -3.259  0.00112 **
s4          -0.38344    0.18505  -2.072  0.03826 *
s5          -0.82098    0.25584  -3.209  0.00133 **
s6          -1.22422    0.21718  -5.637 1.73e-08 ***
s7          -0.18813    0.18029  -1.043  0.29672
s8           0.03922    0.19807   0.198  0.84303
s9          -1.42489    0.22851  -6.236 4.50e-10 ***
s10          0.21511    0.19008   1.132  0.25776
CS2          0.68981    0.10006   6.894 5.42e-12 ***
---
Signif. codes:  0 '***' 0.001 '**' 0.01 '*' 0.05 '.' 0.1 ' ' 1
```

The estimated values of conditional symmetry are stated below. From the estimated values, we can observe that the estimated ratio between upper and lower off-diagonal cells $\hat{\gamma}$ is a constant value of 1.99. For instance, the ratio of $\hat{m}_{21}$–$\hat{m}_{12}$ is a constant value of 1.99, and the ratio of $\hat{m}_{31}$–$\hat{m}_{13}$ is also a value of 1.99. This ratio can also be derived from the estimated coefficient of CS2, $\exp(0.68981) = 1.99$.

$$
\hat{m}_{ij} = \begin{bmatrix} 50 & 21.05 & 26.73 & 34.08 \\ 41.95 & 22 & 14.70 & 41.43 \\ 53.27 & 29.30 & 52 & 12.03 \\ 67.92 & 82.57 & 23.97 & 62 \end{bmatrix}
$$

$$
\hat{\gamma} = \frac{\hat{m}_{21}}{\hat{m}_{12}} = \frac{41.95}{21.05} = 1.99, \quad \hat{\gamma} = \frac{\hat{m}_{31}}{\hat{m}_{13}} = \frac{53.27}{26.73} = 1.99
$$

In general, all the estimated upper diagonal cells in the same diagonal distance to the lower diagonal cells have a ratio of 1.99.

$$
\hat{\gamma} = \frac{\hat{m}_{21}}{\hat{m}_{12}} = \frac{\hat{m}_{31}}{\hat{m}_{13}} = \cdots = \frac{\hat{m}_{32}}{\hat{m}_{23}} = \frac{\hat{m}_{43}}{\hat{m}_{34}} = 1.99
$$

This ratio 1.99 indicates the vision of right eye is 1.99 that of left eye, $\pi_{ij} = 1.99\pi_{ji}$. That is, right eye is almost twice better than the left eye for these 635 patients.

As each of the lower and upper off-diagonal cells in the relative position to the diagonal is a constant ratio, the ratio of the total lower and total upper off-diagonal cells is also a constant term. That is, the subtotal of upper diagonal is a constant term of the subtotal of the lower diagonal, $\Delta_{\text{upper}} = \gamma\Delta_{\text{lower}}$. For the above example, $\hat{\Delta}_{\text{upper}} = 150.02$, $\hat{\Delta}_{\text{lower}} = 298.98$, and $\hat{\gamma} = 1.99$.

4.3 Odds Symmetry Model

The marginal homogeneity symmetry model examines the marginal total symmetry, the complete symmetry is about symmetrical pattern of the off-diagonal individual cells, and the conditional symmetry model looks at the off-diagonal cell probability symmetry with a restriction of a constant term. This section extends the idea of symmetrical pattern, not applying only to all the off-diagonal cells, but by restricting to a smaller set of the off-diagonal cells, called the odds symmetry model. The odds symmetry models display the symmetrical patterns by restricting to specific cells of rows and columns. It comes in two forms, the type I and type II odds symmetry. The odds symmetry models type I and II (OS1 & OS2) are extensions of the conditional symmetry model (Tomizawa 1985).

Odds symmetry (OS) model states that the odds of a group of cells has the same value for the specific row (i) or column (j) for cell (i,j) to cell (j,i). This common value shares the relationship represented by the constant row odds r_i for OS1 and the constant column odds s_j for OS2. These relationships are stated below as two hypotheses for OS1 and OS2, respectively.

$$H_{OS1} : \frac{\pi_{ij}}{\pi_{(i,j+1)}} = \frac{\pi_{ji}}{\pi_{(j+1,i)}} \qquad H_{OS1} : \frac{\pi_{ij}}{\pi_{ji}} = r_i \quad \text{for } i<j$$

$$H_{OS2} : \frac{\pi_{(i-1,j)}}{\pi_{ij}} = \frac{\pi_{(j,i-1)}}{\pi_{ji}} \qquad H_{OS2} : \frac{\pi_{ij}}{\pi_{ji}} = s_j \quad \text{for } i<j$$

The odds symmetry type I model indicates that the odds of the column value in j instead of $j + 1$ in row i in the upper right triangle of the table is equal to the symmetric odds of the row value in j instead of $j + 1$ in column i in the lower left triangle of the table (Tomizawa 1985). Similarly, odds symmetry type II model indicates that the odds of the row value in $i - 1$ instead of i in column j in the upper right triangle of the table is equal to the symmetric odds of the column value in $i - 1$ instead of i in row j in the lower left triangle of the table. This way of explaining OS1 and OS2 could be quite confusing to some readers who are not very mathematically inclined even though it is precisely elucidated. The two symbolic tables below for OS1 and OS2 provide an easier explanation even though it takes a longer explanation, but once it is understood, it is quite easy to refer to the symbolic table to get the idea.

Figure 4.4 depicts the characteristics of OS1 in a symbolic table for a 5×5 doubly classified table. If we sum the probabilities of cell (1,2), cell (1,3), cell (1,4), and cell (1,5) as the first group, and cell (2,1), cell (3,1), cell (4,1) as the second group, the ratio of the first group and the second group is a constant term r_1. This ratio r_1 also applies to the individual odds between the two cells (1,2) and (2,1), the odds between cell (1,3) and cell (3,1), the odds between cell (1,4) and cell (4,1), and the odds between cell (1,5) and cell (5,1). That is, $\pi_{12}/\pi_{21} = \pi_{13}/\pi_{31} = \pi_{14}/\pi_{41} = \pi_{15}/\pi_{51} = (\pi_{12} + \pi_{13} + \pi_{14} + \pi_{15})/(\pi_{21} + \pi_{31} + \pi_{41} + \pi_{51}) = r_1$. In summary, those cells in Fig. 4.4 that are enclosed within the same shape (□,◻,○,◁) have the same constant r_i. For instance, the odds of cell (3,4) and cell (4,3), the odds of cell (3,5) and cell (5,3), and the odds of the total value of cell (3,4) and cell (3,5) to total value of cell (4,3) and cell (5,3) have a constant value of r_3, the four cells enclosed by the shape ○.

Fig. 4.4 Odds symmetry I model

	1	2	3	4	5
1		π_{12}	π_{13}	π_{14}	π_{15}
2	π_{21}		π_{23}	π_{24}	π_{25}
3	π_{31}	π_{32}		π_{34}	π_{35}
4	π_{41}	π_{42}	π_{43}		π_{45}
5	π_{51}	π_{52}	π_{53}	π_{54}	

Similar interpretation is for OS2, as shown in Fig. 4.5 of symbolic table . Those cells with the same shape (□,◇,○,△) show a constant s_j. For instance, the odds of cell (3,1) and cell (1,3) and the odds of cell (3,2) and cell (2,3) have a constant value of s_2, the four cells enclosed by the shape ○.

The nonstandard log-linear models for OS1 and OS2 with their respective factor OS1 and OS2 are stated below. The degrees of freedom for both OS1 and OS2 models is $(a-1)(a-2)$.

$$OS1 : \ln\left(\hat{m}_{ij}\right) = \mu + \lambda^S_{ij} + \lambda^{OS1}_{ij}$$
$$OS2 : \ln\left(\hat{m}_{ij}\right) = \mu + \lambda^S_{ij} + \lambda^{OS2}_{ij}$$

The OS1 and OS2 factor variables can be generated using the following formulation (Lawal 2004).

$$OS1_{ij} = \begin{cases} i & i<j \\ (a-1)+j & i>j \\ 2a-1 & i=j \end{cases}$$

$$OS2_{ij} = \begin{cases} (2a-j) & i<j \\ (a+1)-i & i>j \\ 2a-1 & i=j \end{cases}$$

The following shows the OS1 and OS2 factor variables for 4×4 doubly classified table.

$$OS1 = \begin{bmatrix} 7 & 1 & 1 & 1 \\ 4 & 7 & 2 & 2 \\ 4 & 5 & 7 & 3 \\ 4 & 5 & 6 & 7 \end{bmatrix}, \quad OS2 = \begin{bmatrix} 7 & 3 & 2 & 1 \\ 6 & 7 & 2 & 1 \\ 5 & 5 & 7 & 1 \\ 4 & 4 & 4 & 7 \end{bmatrix}$$

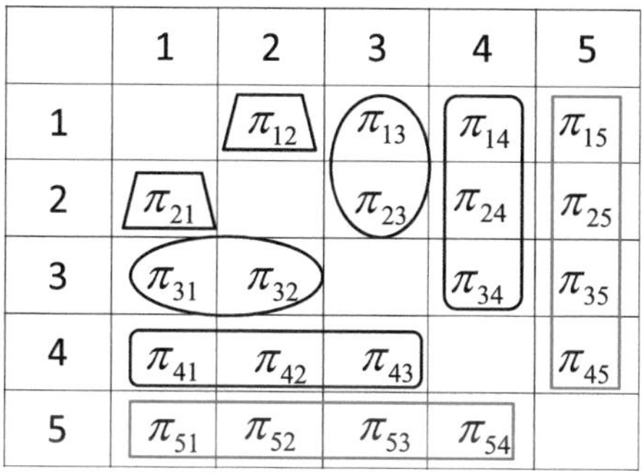

Fig. 4.5 Odds symmetry II model

	1	2	3	4	5
1		π_{12}	π_{13}	π_{14}	π_{15}
2	π_{21}		π_{23}	π_{24}	π_{25}
3	π_{31}	π_{32}		π_{34}	π_{35}
4	π_{41}	π_{42}	π_{43}		π_{45}
5	π_{51}	π_{52}	π_{53}	π_{54}	

A 6 × 6 doubly classified Table 4.5 records the ratings of raters A and B of a school of 916 pupils. Grade of 1 represents the lowest grading, and grade of 6 represents highest rating.

The R syntax to generate odds symmetry I model for the above table is given below. Specifying both factors *s* and os1 in the glm() function generates the OS1 model.

```
OS1Data <- c(50, 10, 22, 6, 23, 11,
             21, 20, 44, 22, 27, 45,
             40, 21, 60, 20, 13, 16,
             15, 11, 21, 40, 22, 13,
             40, 13, 12, 34, 55, 5,
             20, 23, 25, 20, 10, 66)
 s <- c(1, 2, 3, 4, 5, 6,
             2, 7, 8, 9,10,11,
             3, 8,12,13,14,15,
             4, 9,13,16,17,18,
             5,10,14,17,19,20,
             6,11,15,18,20,21)
s <- as.factor(s)
os1<- c(11, 1, 1, 1, 1, 1,
             6,11, 2, 2, 2, 2,
             6, 7,11, 3, 3, 3,
             6, 7, 8,11, 4, 4,
             6, 7, 8, 9,11, 5,
             6, 7, 8, 9,10,11)
 os1 <- as.factor(os1)
OS1<-glm(OS1Data~s+os1,family=poisson)
summary(OS1)
model.summary(OS1)
```

Table 4.5 Rater A × rater B

Rater B	Rater A					
	1	2	3	4	5	6
1	50	10	22	6	23	11
2	21	20	44	22	27	45
3	40	21	60	20	13	16
4	15	11	21	40	22	13
5	40	13	12	34	55	5
6	20	23	25	20	10	66

The estimated values of OSI are given below.

$$
\hat{m}_{ij} =
\begin{bmatrix}
50 & 10.73 & 21.46 & 7.27 & 21.81 & 10.73 \\
20.27 & 20 & 43.54 & 22.11 & 26.8 & 45.55 \\
40.54 & 21.46 & 60 & 18.78 & 11.45 & 18.78 \\
13.73 & 10.89 & 22.22 & 40 & 22.02 & 12.98 \\
41.19 & 13.2 & 13.55 & 33.98 & 55 & 5 \\
20.27 & 22.45 & 22.22 & 20.02 & 10 & 66
\end{bmatrix}
$$

There are five r_i for a 6×6 table. The values of r_1-r_5 are $r = (0.53 \quad 2.03 \quad 0.84 \quad 0.65 \quad 0.50)$.

$$
r_1 = \frac{\hat{m}_{12}}{\hat{m}_{21}} = \frac{10.73}{20.27} = \frac{\hat{m}_{13}}{\hat{m}_{31}} = \frac{21.46}{40.54} = \cdots = \frac{\hat{m}_{16}}{\hat{m}_{61}} = \frac{10.73}{20.27} = 0.53
$$

$$
r_2 = \frac{\hat{m}_{23}}{\hat{m}_{32}} = \frac{43.54}{21.46} = \cdots = \frac{\hat{m}_{26}}{\hat{m}_{62}} = \frac{45.55}{22.45} = 2.03
$$

$$
r_3 = \frac{\hat{m}_{34}}{\hat{m}_{43}} = \frac{18.78}{22.22} = \cdots = \frac{\hat{m}_{36}}{\hat{m}_{63}} = \frac{18.78}{22.22} = 0.84
$$

$$
r_4 = \frac{\hat{m}_{45}}{\hat{m}_{54}} = \frac{22.02}{33.98} = \frac{\hat{m}_{46}}{\hat{m}_{64}} = \frac{12.98}{20.02} = 0.65
$$

$$
r_5 = \frac{\hat{m}_{56}}{\hat{m}_{65}} = \frac{5}{10} = 0.5
$$

The estimated log odds ratios of odds symmetry I have the following property between log odds ratio $\hat{\Phi}_{ij}$ and r_i.

$$
\hat{\Phi}_{ij} - \hat{\Phi}_{ji} =
\begin{cases}
\ln(r_{i+1}) & |i - j| = 1 \\
0 & \text{Otherwise}
\end{cases}
$$

The estimated odds ratios and log odds ratios are printed below.

$$
\hat{\theta}_{ij} =
\begin{bmatrix}
4.60 & 1.09 & 1.50 & 0.40 & 3.45 \\
0.54 & 1.28 & 0.62 & 0.50 & 0.97 \\
1.50 & 0.73 & 5.75 & 0.90 & 0.36 \\
0.40 & 0.50 & 1.39 & 2.94 & 0.15 \\
3.46 & 0.96 & 0.36 & 0.31 & 72.6
\end{bmatrix}
$$

$$
\hat{\Phi}_{ij} =
\begin{bmatrix}
1.53 & 0.08 & 0.40 & -0.91 & 1.24 \\
-0.62 & 0.25 & -0.48 & -0.69 & -0.04 \\
0.40 & -0.32 & 1.75 & -0.10 & -1.02 \\
-0.91 & -0.69 & 0.33 & 1.08 & -1.87 \\
1.24 & -0.04 & -1.02 & -1.18 & 4.28
\end{bmatrix}
$$

The following shows the relationship between log odds ratio $\hat{\Phi}_{ij}$ and r_i.

$$\hat{\Phi}_{12} - \hat{\Phi}_{21} = 0.71 = \ln(r_2) = \ln(2.03)$$

$$\hat{\Phi}_{23} - \hat{\Phi}_{32} = -0.17 = \ln(r_3) = \ln(0.84)$$

$$\hat{\Phi}_{34} - \hat{\Phi}_{43} = -0.43 = \ln(r_4) = \ln(0.65)$$

$$\hat{\Phi}_{45} - \hat{\Phi}_{54} = -0.69 = \ln(r_5) = \ln(0.5)$$

$$\hat{\Phi}_{13} - \hat{\Phi}_{31} = \hat{\Phi}_{14} - \hat{\Phi}_{41} = \cdots = 0$$

The following table shows the rating of raters A and B of another school.

The R syntax to generate odds symmetry II model for the above table is given below. The difference between OS1 and OS2 is that the factor specification of OS1 is s+OS1, which now becomes s+OS2.

```
OS2Data <- c(56, 10, 12, 20, 23, 11,
             21, 23, 8, 18, 7, 35,
             40, 21, 11, 33, 13, 16,
             13, 11, 21, 38, 20, 13,
             43, 13, 27, 34, 56, 5,
             20, 71, 33, 27, 10, 11)
s <- c(1, 2, 3, 4, 5, 6,
       2, 7, 8, 9,10,11,
       3, 8,12,13,14,15,
       4, 9,13,16,17,18,
       5,10,14,17,19,20,
       6,11,15,18,20,21)
s <- as.factor(s)
os2 <- c(11,10, 9, 8, 7, 6,
          5,11, 9, 8, 7, 6,
          4, 4,11, 8, 7, 6,
          3, 3, 3,11, 7, 6,
          2, 2, 2,11, 6,
          1, 1, 1, 1, 1,11)
os2 <- as.factor(os2)
OS2 <-glm(OS2Data~s+os2,family=poisson)
summary(OS2)
model.summary(OS2)
```

The estimated values of OS2 are stated below.

$$\hat{m}_{ij} = \begin{bmatrix} 56 & 10 & 12.84 & 20.20 & 23.10 & 10.29 \\ 21 & 23 & 7.16 & 17.75 & 7 & 35.19 \\ 39.16 & 21.84 & 11 & 33.05 & 14 & 16.27 \\ 12.80 & 11.25 & 20.95 & 38 & 18.9 & 13.28 \\ 42.90 & 13 & 26 & 35.10 & 56 & 4.98 \\ 20.71 & 70.81 & 32.73 & 26.72 & 10.02 & 11 \end{bmatrix}$$

The estimated s_j for odds symmetry II are $s = (\,0.49 \quad 0.53 \quad 1.58 \\ 0.33 \quad 0.48\,)$.

$$s_1 = \frac{\hat{m}_{16}}{\hat{m}_{61}} = \frac{10.29}{20.71} = \frac{\hat{m}_{26}}{\hat{m}_{62}} = \frac{35.19}{70.81} = \cdots = \frac{\hat{m}_{56}}{\hat{m}_{65}} = \frac{4.98}{10.02} = 0.49$$

$$s_2 = \frac{\hat{m}_{15}}{\hat{m}_{51}} = \frac{23.10}{42.90} = \cdots = \frac{\hat{m}_{45}}{\hat{m}_{54}} = \frac{18.90}{35.10} = 0.53$$

$$s_3 = \frac{\hat{m}_{14}}{\hat{m}_{41}} = \frac{20.20}{12.80} = \cdots = \frac{\hat{m}_{34}}{\hat{m}_{43}} = \frac{33.05}{20.95} = 1.58$$

$$s_4 = \frac{\hat{m}_{13}}{\hat{m}_{31}} = \frac{12,84}{39.16} = \frac{\hat{m}_{23}}{\hat{m}_{32}} = \frac{7.16}{21.84} = 0.33$$

$$s_5 = \frac{\hat{m}_{12}}{\hat{m}_{21}} = \frac{10}{21} = 0.48$$

Similar to odds symmetry I model, the estimated log odds ratios of odds symmetry II have the following property between log odds ratio $\hat{\Phi}_{ij}$ and s_j.

$$\hat{\Phi}_{ij} - \hat{\Phi}_{ji} = \begin{cases} \ln(s_{I-i}) & |i-j| = 1 \\ 0 & \text{Otherwise} \end{cases}$$

The estimated odds ratios and log odds ratios are stated below.

$$\hat{\theta}_{ij} = \begin{bmatrix} 6.13 & 0.24 & 1.58 & 0.34 & 11.29 \\ 0.51 & 1.62 & 1.21 & 1.07 & 0.23 \\ 1.58 & 3.70 & 0.60 & 1.17 & 0.60 \\ 0.34 & 1.07 & 0.74 & 3.21 & 0.13 \\ 11.28 & 0.23 & 0.60 & 0.24 & 12.34 \end{bmatrix}$$

$$\hat{\Phi}_{ij} = \begin{bmatrix} 1.81 & -1.42 & 0.45 & -1.06 & 2.42 \\ -0.67 & 0.48 & 0.19 & 0.07 & -1.46 \\ 0.45 & 1.31 & -0.50 & 0.16 & -0.50 \\ -1.06 & 0.07 & -0.30 & 1.17 & -2.07 \\ 2.42 & -1.46 & -0.50 & -1.45 & 2.51 \end{bmatrix}$$

Table 4.6 Rater A × rater B

Rater B	Rater A					
	1	2	3	4	5	6
1	56	10	12	20	23	11
2	21	23	8	18	7	35
3	40	21	11	33	13	16
4	13	11	21	38	20	13
5	43	13	27	34	56	5
6	20	71	33	27	10	11

The relationship between $\hat{\Phi}_{ij}$ and s_j for odds symmetry II model in Table 4.6 is stated below.

$$\hat{\Phi}_{12} - \hat{\Phi}_{21} = -0.74 = \ln(c_5) = \ln(0.48)$$

$$\hat{\Phi}_{23} - \hat{\Phi}_{32} = -1.12 = \ln(c_4) = \ln(0.33)$$

$$\hat{\Phi}_{34} - \hat{\Phi}_{43} = 0.46 = \ln(c_3) = \ln(1.58)$$

$$\hat{\Phi}_{45} - \hat{\Phi}_{54} = -0.62 = \ln(c_2) = \ln(0.53)$$

$$\hat{\Phi}_{13} - \hat{\Phi}_{31} = \hat{\Phi}_{14} - \hat{\Phi}_{41} = \cdots = 0$$

4.3.1 Relationships of Complete Symmetry, Conditional Symmetry, Odds Symmetry I and II

The relationships of complete symmetry, conditional symmetry, odds symmetry I and II models are depicted in Fig. 4.6. When $\gamma = 1$, conditional symmetry model reduces to complete symmetry model. This means that when the probability of all the cells in upper diagonal to all the cells in lower diagonal is equally likely, conditional symmetry reduces to complete symmetry model. Odds symmetry model I reduces to complete symmetry model and to conditional symmetry when $r_i = 1$

Fig. 4.6 Relationships of complete symmetry, conditional symmetry, odds symmetry I and II

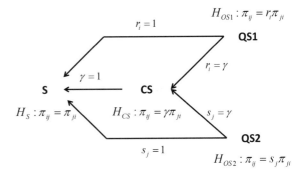

and $r_i = \gamma$, respectively. Similarly, odds symmetry model II reduces to complete symmetry model and to conditional symmetry when $s_j = 1$ and $s_j = \gamma$, respectively. These relationships show the hierarchical links of the four doubly classified models. CS is a generalization of S, and OSI and OSII are generalization models of CS and S. The hierarchical nature of asymmetry models will be discussed and summarized in Chap. 10 when more asymmetry models are introduced.

4.4 Diagonal Parameters Symmetry Model

The odds symmetry I and II exhibit symmetrical odds according to the position of row and column. Goodman (1979a) and Agresti (1983) proposed the diagonal parameters symmetry (DPS) model which the odds exhibit a symmetrical pattern according to the position of cells from the diagonal axis. The hypothesis of the diagonal parameters symmetry model is as follows:

$$H_{\text{DPS}} : \pi_{ij} = \pi_{ji}\delta_k \quad \text{for} \ i > j \ k = i - j$$

where δ_k is the odds of observation falls in cell (i,j) instead of cell (j,i), and k is the distance from the diagonal axis where $k = i - j$. The following symbolic table shows the characteristics of odds of diagonal parameters symmetry model. The value of odds are the same for cells at the same position from the diagonal. The value of odds from different positions differ. As such, the values of δ_k differ (Fig. 4.7).

The nonstandard log-linear model of DPS is specified below with factor variables S and D. The specifications of S and D for 4×4 table are shown below.

$$\ln\left(\hat{m}_{ij}\right) = \mu + \lambda_{ij}^S + \lambda_{ij}^D$$

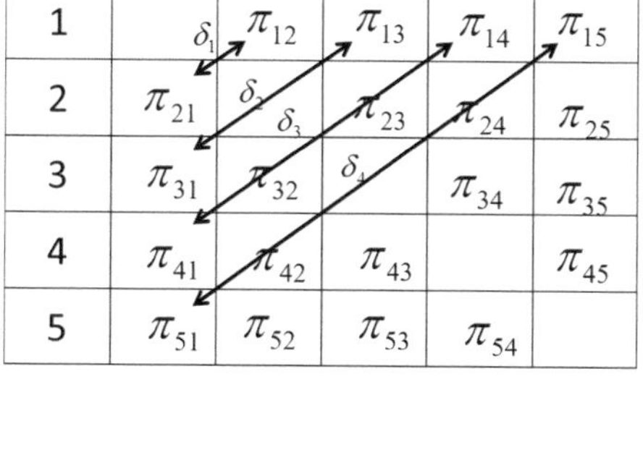

Fig. 4.7 Diagonal parameters symmetry model: symmetry in δ_k from the diagonal

$$S = \begin{bmatrix} 1 & 2 & 3 & 4 \\ 2 & 5 & 6 & 7 \\ 3 & 6 & 8 & 9 \\ 4 & 7 & 9 & 10 \end{bmatrix}, \quad D = \begin{bmatrix} 7 & 1 & 2 & 3 \\ 4 & 7 & 1 & 2 \\ 5 & 4 & 7 & 1 \\ 6 & 5 & 4 & 7 \end{bmatrix}$$

The following table tabulates the ratings of two raters 1 and 2 on a group of examinees sat for a test. Both raters rate them into four categories: excellent, good, average, and below average.

The factors S and D for 5×5 doubly classified table are as follows:

$$S = \begin{bmatrix} 1 & 2 & 3 & 4 & 5 \\ 2 & 6 & 7 & 8 & 9 \\ 3 & 7 & 10 & 11 & 12 \\ 4 & 8 & 11 & 13 & 14 \\ 5 & 9 & 12 & 14 & 15 \end{bmatrix}, \quad D = \begin{bmatrix} 9 & 1 & 2 & 3 & 4 \\ 5 & 9 & 1 & 2 & 3 \\ 6 & 5 & 9 & 1 & 2 \\ 7 & 6 & 5 & 9 & 1 \\ 8 & 7 & 6 & 5 & 9 \end{bmatrix}$$

The R syntax to generate the diagonal parameters symmetry model for Table 4.7 is printed below.

```
Rater <- c(30, 20, 32, 11, 53,
           31, 40, 31, 12,  9,
           15, 64, 32,  9, 22,
           35,  7, 16, 78, 15,
           25, 20, 15, 32, 60)
s <- c( 1, 2,  3,  4,   5,
        2, 6,  7,  8,   9,
        3, 7, 10, 11, 12,
        4, 8, 11, 13, 14,
        5, 9, 12, 14, 15)
s <- as.factor(s)
d <- c( 9, 1, 2, 3, 4,
        5, 9, 1, 2, 3,
        6, 5, 9, 1, 2,
        7, 6, 5, 9, 1,
```

Table 4.7 Rating of two raters

Rater 2	Rater 1				
	Excellent	Good	Average	Below average	Bad
Excellent	30	20	32	11	53
Good	31	40	31	12	9
Average	15	64	32	9	22
Below Average	35	7	16	78	15
Bad	25	20	15	32	60

```
     8, 7, 6, 5, 9)
d <- as.factor(d)
DPSM<-glm(Rater~s+d,family=poisson)
summary(DPSM)
model.summary(DPSM)
```

The estimated values for diagonal parameter model are printed below.

$$\hat{m}_{ij} = \begin{bmatrix} 30 & 17.55 & 30.12 & 12.27 & 53 \\ 33.45 & 40 & 32.68 & 12.17 & 7.73 \\ 16.88 & 62.32 & 32 & 8.6 & 23.71 \\ 33.73 & 6.83 & 16.4 & 78 & 16.17 \\ 25 & 21.27 & 13.29 & 30.83 & 60 \end{bmatrix}$$

The estimated δ_k are stated below.

$$\hat{\delta}_1 = \frac{\hat{m}_{21}}{\hat{m}_{12}} = 0.52, \quad \hat{\delta}_2 = \frac{\hat{m}_{31}}{\hat{m}_{13}} = 1.78,$$

$$\hat{\delta}_3 = \frac{\hat{m}_{41}}{\hat{m}_{14}} = 0.36, \quad \hat{\delta}_4 = \frac{\hat{m}_{51}}{\hat{m}_{15}} = 2.12$$

As mentioned above, the odds of those cells with the same distance away from the diagonal are with the same value, as shown below.

$$\hat{\delta}_1 = \frac{\hat{m}_{21}}{\hat{m}_{12}} = \frac{\hat{m}_{23}}{\hat{m}_{32}} = \frac{\hat{m}_{34}}{\hat{m}_{43}} = \frac{\hat{m}_{45}}{\hat{m}_{54}},$$

$$\hat{\delta}_2 = \frac{\hat{m}_{31}}{\hat{m}_{13}} = \frac{\hat{m}_{24}}{\hat{m}_{42}} = \frac{\hat{m}_{35}}{\hat{m}_{53}}, \quad \text{and} \quad \hat{\delta}_3 = \frac{\hat{m}_{41}}{\hat{m}_{14}} = \frac{\hat{m}_{25}}{\hat{m}_{52}}.$$

Another properties of diagonal parameters symmetry model are that the difference between diagonal cells in log term in the opposite diagonal is a constant term. For instance, for a 4×4 doubly classified table, the expected log difference between paired cell of (1,2) and (2,1), paired cell of (2,3) and (3,2), and paired cell of (3,4) and (4,3) is a common constant term $\ln(\delta_1)$. These are cells one position away from the diagonal, i.e., $k = 1 = i - j$. We get another common constant term $\ln(\delta_2)$ for cells that are two positions away from the diagonal. In essence, the relationship between off-diagonal cells is expressed in relationship of odds or ln (odds) as far as their distance to the diagonal is concerned.

$$\ln(\hat{m}_{ji}) - \ln(\hat{m}_{ij}) = \hat{\tau}_{i-j}$$
$$\text{where } \tau_{i-j} = \ln(\delta_{i-j})$$

$$\hat{\tau}_1 = \ln(\hat{m}_{12}) - \ln(\hat{m}_{21}) = \exp\left(\hat{\delta}_1\right) = 1.67$$

$$\hat{\tau}_2 = \ln(\hat{m}_{13}) - \ln(\hat{m}_{31}) = \exp\left(\hat{\delta}_2\right) = 5.93$$

$$\hat{\tau}_3 = \ln(\hat{m}_{14}) - \ln(\hat{m}_{41}) = \exp\left(\hat{\delta}_3\right) = 1.43$$

$$\hat{\tau}_4 = \ln(\hat{m}_{15}) - \ln(\hat{m}_{51}) = \exp\left(\hat{\delta}_4\right) = 8.33$$

Similar to $\hat{\delta}$ that stated above,

$$\hat{\tau}_1 = \ln(\hat{m}_{23}) - \ln(\hat{m}_{32}) = \ln(\hat{m}_{34}) - \ln(\hat{m}_{43}) = \ln(\hat{m}_{45}) - \ln(\hat{m}_{54}),$$
$$\hat{\tau}_2 = \ln(\hat{m}_{24}) - \ln(\hat{m}_{42}) = \ln(\hat{m}_{35}) - \ln(\hat{m}_{53}), \quad \text{and} \quad \hat{\tau}_3 = \ln(\hat{m}_{25}) - \ln(\hat{m}_{52})$$

4.4.1 Relationships of Complete Symmetry, Conditional Symmetry, and Diagonal Parameters Symmetry Models

Similar to odds symmetry model, diagonal parameters symmetry model is also an extension of the conditional symmetry model. When the following condition for δ_k satisfies, DPS becomes the conditional symmetry model where $\pi_{ij} = \gamma\pi_{ji}$. That is, when $\delta_k = \gamma$, δ_k reduces to a constant term γ.

$$\delta_1 = \delta_2 = \cdots = \delta_{a-1} = \gamma$$

For conditional symmetry model, it is a direct single relationship of γ between the upper and lower diagonal cells, whereas DPS further distinguishes those cells that are nearer to the diagonal and those further away with different odds values of δ_k. When all $\delta_k = 1$, the meaning of δ_k becomes worthless, and thus, it reduces to a complete symmetry model. Figure 4.8 depicts the nesting relationship of these three models.

Fig. 4.8 Relationships of complete symmetry, conditional symmetry, and diagonal parameters symmetry models

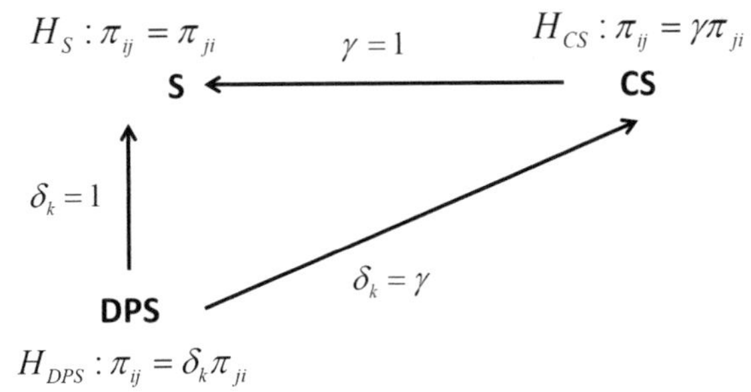

4.5 Linear Diagonal Parameters Symmetry Model

The diagonal parameters symmetry model exhibits the characteristics that cells in reference to the diagonal are symmetrical in their odds. Similarly, the linear diagonal parameters symmetry model also indicates such a pattern of association in their odds in reference to the diagonal. The difference between the two models is that the latter exhibits a linear relationship in their odds as shown in Fig. 4.9. As such, Agresti (1983) considers the linear diagonal parameters symmetry (LDSP) model as a simpler model than the diagonal parameters symmetry model.

$$H_{\text{LDPS}} : \pi_{ij} = \pi_{ji}\delta^{j-i} \quad \text{for } j \geq i$$

The symbolic table below describes the characteristics of the linear diagonal parameters model. Those cells that are one position away from the diagonal have an odds of δ, two positions away have an odds of δ^2, and so on (Fig. 4.9).

The nonstandard log-linear model specification of linear diagonal parameters symmetry model is specified as follows:

$$\ln\left(\hat{m}_{ij}\right) = \mu + \lambda_{ij}^{S} + \tau\lambda_{ij}^{F}$$

$$S = \begin{bmatrix} 1 & 2 & 3 & 4 \\ 2 & 5 & 6 & 7 \\ 3 & 6 & 8 & 9 \\ 4 & 7 & 9 & 10 \end{bmatrix}, \quad F = \begin{bmatrix} 1 & 2 & 3 & 4 \\ 1 & 1 & 2 & 3 \\ 1 & 1 & 1 & 2 \\ 1 & 1 & 1 & 1 \end{bmatrix}$$

The model is based on $(a+1)(a-2)/2$ degrees of freedom. Noted that S is a factor variable, whereas F is not a factor variable but a rank variable. This rank variable F specifies the linear pattern of odds. The estimated scalar τ for F represents the distance in logarithm term away from a one-step off-diagonal cell. It is a constant term that measures the differences of the one step away from the diagonal to the two steps, three steps, and so on.

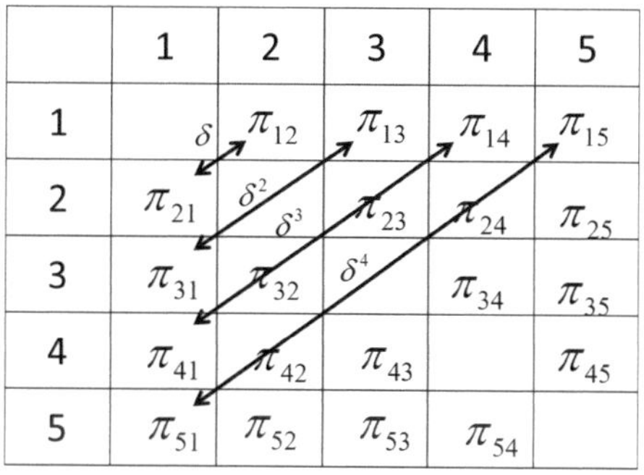

Fig. 4.9 Linear diagonal parameters symmetry model

Table 4.8 Current residence
and residence at age 21

Residence at age 21	Current residence			
	Toa Payoh	Bishan	Bedok	Tampines
Toa Payoh	520	167	133	57
Bishan	233	512	423	88
Bedok	109	370	172	205
Tampines	45	73	180	500

Table 4.8 tabulates the 3787 Singaporeans—the town they lived in at age 21 and their current residence.

The R syntax to generate the linear diagonal parameters symmetry model for the above table is as follows:

```
Residence <- c(520, 167, 133, 57,
               233, 512, 423, 88,
               109, 370, 172, 205,
                45,  73, 180, 500)
s<-c( 1, 2, 3, 4,
      2, 5, 6, 7,
      3, 6, 8, 9,
      4, 7, 9,10)
s <- as.factor(s)
F <- c(1, 2, 3, 4,
       1, 1, 2, 3,
       1, 1, 1, 2,
       1, 1, 1, 1)
LQPS<-glm(Residence~s+F,family=poisson)
summary(LQPS)
model.summary(LQPS)
```

The estimated coefficients of the linear diagonal parameters are printed below.

```
> summary(LQPS)

Call:
glm(formula = Residence ~ s + F, family = poisson)

Deviance Residuals:
      1        2        3        4        5        6        7        8
 0.0000  -2.8200   0.4016   0.1755   2.7237   0.0000   0.7046   0.2758
      9       10       11       12       13       14       15       16
-0.4325  -0.7354   0.0000   0.4653  -0.1941  -0.2964  -0.4853   0.0000

Coefficients:
            Estimate Std. Error z value Pr(>|z|)
(Intercept)  6.19241    0.05386 114.982  < 2e-16 ***
s2          -0.98669    0.06843 -14.419  < 2e-16 ***
s3          -1.52134    0.08459 -17.984  < 2e-16 ***
s4          -2.41837    0.11979 -20.189  < 2e-16 ***
s5          -0.01550    0.06226  -0.249   0.8033
s6          -0.30233    0.05868  -5.152 2.58e-07 ***
s7          -1.92888    0.09610 -20.072  < 2e-16 ***
s8          -1.10633    0.08796 -12.578  < 2e-16 ***
s9          -1.02491    0.06914 -14.824  < 2e-16 ***
s10         -0.03922    0.06263  -0.626   0.5312
F            0.06142    0.03126   1.965   0.0495 *
---
Signif. codes:  0 '***' 0.001 '**' 0.01 '*' 0.05 '.' 0.1 ' ' 1
```

The estimated values and the odds ratios are shown below. The odds ratios show symmetrical pattern. For instance, $\hat{\theta}_{12} = \hat{\theta}_{21} = 1.28$ and $\hat{\theta}_{13} = \hat{\theta}_{31} = 0.48$.

$$\hat{m}_{ij} = \begin{bmatrix} 520 & 206.14 & 128.42 & 55.69 \\ 193.86 & 512 & 408.67 & 85.44 \\ 113.58 & 384.33 & 172 & 198.41 \\ 46.31 & 75.56 & 186.59 & 500 \end{bmatrix}$$

$$\hat{\theta}_{ij} = \begin{bmatrix} 6.66 & 1.28 & 0.48 \\ 1.28 & 0.56 & 5.52 \\ 0.48 & 5.52 & 2.32 \end{bmatrix}$$

As the odds of the linear DPS have the linear incremental characteristics with regard to the position of the cell, the further away of the cell position from the diagonal, the larger is the log odds. The value of log odds is a constant term in reference to the position of the cell to the diagonal. The estimated τ for the F variable is 0.061 and $\exp(\tau) = 1.0629$. The odd of $\hat{m}_{12}/\hat{m}_{21}$ is equal to $\exp(\tau)$, and the odds of $\hat{m}_{13}/\hat{m}_{31}$ is equal to $\exp(\tau)^2$. The value of odds depends on the position of the cell from the diagonal is the main characteristics of linear DPS. In general, the log odds is a constant term, as shown below, depending on the cell position.

$$\ln\left(\frac{\pi_{ij}}{\pi_{ji}}\right) = \tau(j - i) \quad \text{for } i < j$$

where $\tau = \ln(\delta)$.

The following shows the details of the relationships of odds.

$$\tau = 0.061$$
$$\exp(\tau) = \exp(0.061) = 1.0629$$
$$\frac{\hat{m}_{12}}{\hat{m}_{21}} = \frac{206.14}{193.86} = \frac{\hat{m}_{23}}{\hat{m}_{32}} = \frac{408.67}{384.33} = \frac{\hat{m}_{34}}{\hat{m}_{43}} = \frac{198.41}{186.59} = 1.06$$

$$\exp(\tau)^2 = \exp(0.061)^2 = 1.13$$
$$\frac{\hat{m}_{13}}{\hat{m}_{31}} = \frac{128.42}{113.58} = \frac{\hat{m}_{24}}{\hat{m}_{42}} = \frac{85.44}{75.56} = 1.13$$

$$\exp(\tau)^3 = \exp(0.061)^3 = 1.20$$
$$\frac{\hat{m}_{14}}{\hat{m}_{41}} = \frac{55.69}{46.31} = 1.20$$

The following show the relationships between log odds and τ.

$$\ln\left(\frac{\hat{m}_{12}}{\hat{m}_{21}}\right) = \ln\left(\frac{206.14}{193.86}\right) = 0.061 = (2-1) \times 0.061$$

$$\ln\left(\frac{\hat{m}_{13}}{\hat{m}_{31}}\right) = \ln\left(\frac{128.42}{113.58}\right) = 0.122 = (3-1) \times 0.061$$

$$\ln\left(\frac{\hat{m}_{14}}{\hat{m}_{41}}\right) = \ln\left(\frac{55.69}{46.31}\right) = 0.184 = (4-1) \times 0.061$$

$$\ln\left(\frac{\hat{m}_{23}}{\hat{m}_{32}}\right) = \ln\left(\frac{408.67}{384.33}\right) = 0.061 = (3-2) \times 0.061$$

$$\ln\left(\frac{\hat{m}_{24}}{\hat{m}_{42}}\right) = \ln\left(\frac{85.44}{75.56}\right) = 0.122 = (4-2) \times 0.061$$

$$\ln\left(\frac{\hat{m}_{34}}{\hat{m}_{43}}\right) = \ln\left(\frac{198.41}{186.59}\right) = 0.061 = (4-3) \times 0.061$$

$$\tau = 0.061$$

4.6 Quasi Symmetry Model

While diagonal parameters symmetry model and linear diagonal parameters symmetry model describe the symmetrical pattern of odds of the off-diagonal cells of a table, quasi symmetry model uses odds ratios to depict the relationship of the off-diagonal cells of an odds ratios table. Quasi-symmetry (QS) model exhibits the symmetry of cross product ratios (Goodman 1979b). The specification of the quasi symmetry is as follows:

$$\pi_{ij} = \alpha_i \beta_j \psi_{ij} \quad \text{for } i = 1, 2, \ldots, a; \quad j = 1, 2, \ldots, a$$

where π_{ij} represents the probability that an observation falls in the cell in row i and column j, α_i and β_j are nonnegative parameters pertain to the row marginal and the column marginal in an $a \times a$ table, and $\psi_{ij} = \psi_{ji}$ (Bishop et al. 1975; Caussinus 1965; Tomizawa 1991). The model is based on $(a-1)(a-2)/2$ degrees of freedom. The hypothesis of QS is stated as follows:

$$H_{QS} : \pi_{ij} = \phi_j \pi_{ji}$$

If we let the local odd ratio for cell (i,s), (i,t), (j,s), and (j,t) as $\theta_{(i<j;s<t)} = \pi_{is}\pi_{jt}/\pi_{js}\pi_{it}$, symmetry odd ratio occurs when $\theta_{(i<j;s<t)} = \theta_{(s<t;i<j)}$. The QS model can be expressed as

$$\theta_{(i<j;s<t)} = \theta_{(s<t;i<j)} \quad \text{for } 1 \leq i < j \leq a; \quad 1 \leq s < t \leq a$$

The main characteristic of quasi symmetry is described in the symbolic table below. Figure 4.10 shows the symmetry of local odds ratio of a 4 × 4 odds ratios table. The circle with the same color has the same odds ratios. For instance, the odds ratio of θ_{12} is the same as θ_{21}. Similarly, $\theta_{13} = \theta_{31}$, $\theta_{14} = \theta_{41}$, and so on. This local odds ratio table is derived from a 5 × 5 doubly classified table. As four adjacent cells with 2 cells in adjacent row and 2 cells in adjacent column from a doubly classified table derive one local odds ratio, a 5 × 5 doubly classified table ends up with a 4 × 4 local odds ratio table. The quasi-symmetry (QS) model is characterized by the symmetry in these local odds ratios. As the local odds ratio represents the relationships of four cells in adjacent location, the values of these local odds ratio depict the relationship of changes from one adjacent cell to the next.

The nonstandard log-linear formulation of quasi symmetry model is specified as follows:

$$\ln\left(\hat{m}_{ij}\right) = \mu + \lambda_i^R + \lambda_j^C + \lambda_{ij}^S$$

where the three factors S, R, and C are specified below for 4 × 4 doubly classified table.

$$S = \begin{bmatrix} 1 & 2 & 3 & 4 \\ 2 & 5 & 6 & 7 \\ 3 & 6 & 8 & 9 \\ 4 & 7 & 9 & 10 \end{bmatrix}, \quad R = \begin{bmatrix} 1 & 2 & 3 & 4 \\ 1 & 2 & 3 & 4 \\ 1 & 2 & 3 & 4 \\ 1 & 2 & 3 & 4 \end{bmatrix}, \quad C = \begin{bmatrix} 1 & 1 & 1 & 1 \\ 2 & 2 & 2 & 2 \\ 3 & 3 & 3 & 3 \\ 4 & 4 & 4 & 4 \end{bmatrix}$$

129 students were rated by two teachers A and B. Student gets a score of 1 represents well above average, 2 above average, 3 below average, and 4 well below average. The cross-tabulation of the two teachers is summarized in Table 4.9.

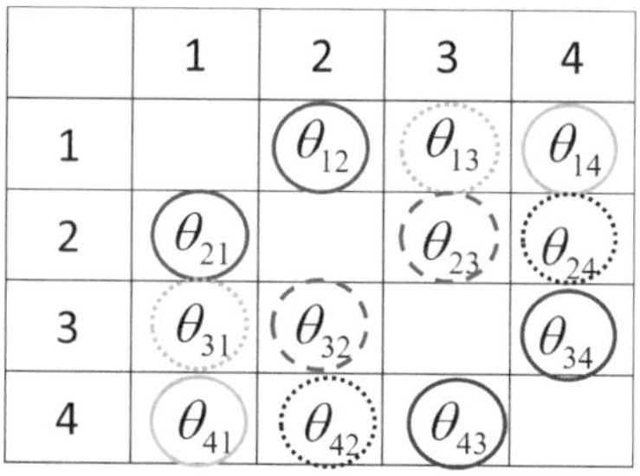

Fig. 4.10 Quasi symmetry model: local odds ratio symmetry

Table 4.9 Rating of teacher A and teacher B

Teacher B	Teacher A			
	1	2	3	4
1	34	5	6	7
2	7	3	6	9
3	6	4	6	5
4	7	6	6	12

The following R syntax generates the quasi symmetry model. The factors r and c stated below show a two-step procedure to create them. First, create a vector using c(), followed by as.factor() function to turn the vector into a factor variable. A shorter way to create r and c factors is to use function gl(). The command gl(4,4) generates a factor with 4 levels and 4 replications, the r factor, and gl(4,1, length = 16) specifies a factor that produces c factor.

```
QS <- c(34, 5, 6,  7,
         7, 3, 6,  9,
         6, 4, 6,  5,
         7, 6, 6, 12)
s<-c( 1, 2, 3, 4,
      2, 5, 6, 7,
      3, 6, 8, 9,
      4, 7, 9,10)
s <- as.factor(s)
r <- c( 1, 1, 1, 1,
        2, 2, 2, 2,
        3, 3, 3, 3,
        4, 4, 4, 4)
r <- as.factor(r)
# Alternatively, r <- gl(4,4)
c <- c( 1, 2, 3, 4,
        1, 2, 3, 4,
        1, 2, 3, 4,
        1, 2, 3, 4)
# Alternatively, c <- gl(4,1, length=16)
QS<-glm(QS~s+r+c, family=poisson)
summary(QS)
model.summary(QS)
```

The estimated values and odds ratios of QS model are as follows:

$$
\hat{m}_{ij} =
\begin{bmatrix}
34 & 4.89 & 6.19 & 6.92 \\
7.11 & 3 & 6.08 & 8.81 \\
5.81 & 3.92 & 6 & 5.27 \\
7.08 & 6.19 & 5.73 & 12
\end{bmatrix}
$$

$$
\hat{\theta}_{ij} =
\begin{bmatrix}
2.93 & 1.60 & 1.30 \\
1.60 & 0.76 & 0.61 \\
1.30 & 0.61 & 2.38
\end{bmatrix}
$$

The estimated odds ratio above shows symmetry in odds ratio. For instance, $\hat{\theta}_{13} = \hat{\theta}_{31} = 1.30$, $\hat{\theta}_{23} = \hat{\theta}_{32} = 0.61$ and $\hat{\theta}_{12} = \hat{\theta}_{21} = 1.60$.

Fig. 4.11 Relationships of complete symmetry, conditional symmetry, and quasi symmetry models

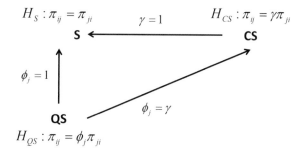

4.6.1 Relationship of Quasi Symmetry, Conditional Symmetry, and Complete Symmetry Models

Quasi symmetry (QS), conditional symmetry (CS), and complete symmetry (S) are hierarchically associated. When $\phi_j = 1$, QS reduces to S, when $\phi_j = \gamma$, QS reduces to CS, and when $\gamma = 1$, CS reduces to S. Complete symmetry (S) implies marginal symmetry and quasi-symmetry; i.e., if we identify a model that is complete symmetry, marginal symmetry and quasi-symmetry are implied. Relating the three models, quasi symmetry model, complete symmetry model, and marginal homogeneity model, Caussinus (1965) gave the theorem that the complete symmetry model holds if and only if both the quasi symmetry and the marginal homogeneity models hold (Bishop et al. 1975; Tomizawa and Tahata 2007). However, when complete symmetry is absent, one could have either marginal or quasi-symmetry but not both (Bhapkar 1979). The following depicts what have described (Fig. 4.11).

4.7 Quasi Diagonal Parameters Symmetry Model

A further decomposition of diagonal parameters symmetry (DPS) model is the quasi-diagonal parameters symmetry (QDPS) model. QDPS is also named as diagonals parameter skew symmetry model (Yamagushi 1990). While DPS shows symmetric pattern in odds in reference to the position from diagonal, and QS exhibits symmetric pattern in terms of θ_{ij}, QDPS further restricts to a group of cells, using θ_{ij} and $\ln\theta_{ij}$ to describe the relationships of cells.

One of the main features of QDPS is that log difference of the expected local odds ratio of cell (i,j) and cell (j,i) has a constant value and the values vary depending on the position of the cell from the diagonal. State in another way, the difference of $\ln\theta_{ij}$ and $\ln\theta_{ji}$ is a constant term in reference to the diagonal. While $\ln\hat{\theta}_{ij} = \hat{\Phi}_{ij}$ represents the estimated local odd ratio from the log odds ratio table of cell (i,j), QDPS has the property that $\hat{\Phi}_{ij} - \hat{\Phi}_{ji}$ is a constant term for cells with the same $k = j - i$. Equivalently, the ratio of θ_{ij} and θ_{ji} is a constant term for the same k.

Fig. 4.12 Quasi diagonal
parameters symmetry

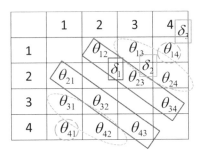

The symbolic table of Fig. 4.12 describes this characteristic of QDPS. This is a
symbolic table of odds ratios. Those cells enclosed by the same shape ($\square;\because;\circlearrowright$)
belong to the same group of cells that gives the same statistical relationship in terms
of log difference in local odds ratio. For instance, the equality of ln
$(\theta_{13}) - \ln(\theta_{31}) = \theta_{24} - \ln(\theta_{42}) = \ln(\theta_{13} + \theta_{24}) - \ln(\theta_{31} + \theta_{42})$ shows the common
relationship of these cells enclosed within the shape $\because$. In essence, QDPS model
displays the relationship of local odds ratio for group of cells taking into account the
distance from the diagonal. The hypothesis of DPS is shown below.

$$H_{\text{QDPS}} : \pi_{ij} = \phi_j \pi_{ji} \delta_k \quad \text{for} \ i < j \ \ k = j - i \ \ k = 1, 2, \ldots, I$$

The nonstandard log-linear form for QDPS is as follows:

$$\ln(\hat{m}_{ij}) = \mu + \lambda_i^R + \lambda_j^C + \lambda_{ij}^S + \lambda_{ij}^D$$

$$\text{where} \quad D_{ij} = \begin{cases} |i - j| & i < j \\ |i - j| + a - 1 & i > j, \\ 2a - 1 & i = j \end{cases}$$

For 4×4 table, factors R, C, S and D are as follows:

$$R = \begin{bmatrix} 1 & 2 & 3 & 4 \\ 1 & 2 & 3 & 4 \\ 1 & 2 & 3 & 4 \\ 1 & 2 & 3 & 4 \end{bmatrix}, \quad C = \begin{bmatrix} 1 & 1 & 1 & 1 \\ 2 & 2 & 2 & 2 \\ 3 & 3 & 3 & 3 \\ 4 & 4 & 4 & 4 \end{bmatrix},$$

$$S = \begin{bmatrix} 1 & 2 & 3 & 4 \\ 2 & 5 & 6 & 7 \\ 3 & 6 & 8 & 9 \\ 4 & 7 & 9 & 10 \end{bmatrix}, \quad \text{and} \ D = \begin{bmatrix} 7 & 1 & 2 & 3 \\ 4 & 7 & 1 & 2 \\ 5 & 4 & 7 & 1 \\ 6 & 5 & 4 & 7 \end{bmatrix}.$$

Table 4.10 tabulates 606 assessments of two raters A and B.
The R codes to generate quasi DPS for the above table are given below.

Table 4.10 Rater A ×
Rater B

Rater A	Rater B				
	1	2	3	4	5
1	30	25	32	11	11
2	21	40	60	12	23
3	15	32	32	9	15
4	35	11	16	32	9
5	28	20	15	12	60

```
Sim <- c( 30, 25, 32, 11, 11,
          21, 40, 60, 12, 23,
          15, 32, 32,  9, 15,
          35, 11, 16, 32,  9,
          28, 20, 15,  12, 60)
s <- c( 1, 2,  3,  4,  5,
        2, 6,  7,  8,  9,
        3, 7, 10, 11, 12,
        4, 8, 11, 13, 14,
        5, 9, 12, 14, 15)
s <- as.factor(s)
r <- c( 1, 1, 1, 1, 1,
        2, 2, 2, 2, 2,
        3, 3, 3, 3, 3,
        4, 4, 4, 4, 4,
        5, 5, 5, 5, 5)
r <- as.factor(r)
c <160;c( 1, 2, 3, 4, 5,
          1, 2, 3, 4, 5,
          1, 2, 3, 4, 5,
          1, 2, 3, 4, 5,
          1, 2, 3, 4, 5)
c <- as.factor(c)
d <- c( 9, 1, 2, 3, 4,
        5, 9, 1, 2, 3,
        6, 5, 9, 1, 2,
        7, 6, 5, 9, 1,
        8, 7, 6, 5, 9)
d <- as.factor(d)
QDPS<-glm(Sim~r+c+s+d,family=poisson)
summary(QDPS)
model.summary(QDPS)
```

```
a <- nrow(Fitted.QDPS)
LocalOdds<-matrix(data=NA,nrow=a-1,ncol=a-1)
for (i in 1:a-1) {
      for (j in 1:a-1) {
          LocalOdds[i,j] - (Fitted.QDPS[i,j] * Fitted.QDPS[i+1,j+1]) /
                           (Fitted.QDPS[i+1,j] * Fitted.QDPS[i,j+1])
      }
}
rownames(LocalOdds) <- c(1:a-1)
colnames(LocalOdds) <- c(1:a-1)
LocalOdds
```

The estimated values, odds ratios, and log odds ratios of QDPS are printed
below.

$$
\hat{m}_{ij} =
\begin{bmatrix}
30 & 21.73 & 31.57 & 14.70 & 11 \\
24.27 & 40 & 59.93 & 12.50 & 19.30 \\
15.43 & 32.07 & 32 & 8.57 & 14.93 \\
31.30 & 10.50 & 16.43 & 32 & 12.77 \\
28 & 23.70 & 15.07 & 8.23 & 60
\end{bmatrix}
$$

$$
\hat{\theta}_{ij} =
\begin{bmatrix}
2.28 & 1.03 & 0.45 & 2.06 \\
1.26 & 0.67 & 1.28 & 1.13 \\
0.16 & 1.57 & 7.27 & 1.13 \\
2.52 & 0.41 & 0.28 & 19.27
\end{bmatrix}
$$

$$
\hat{\Phi}_{ij} =
\begin{bmatrix}
0.82 & 0.03 & -0.80 & 0.72 \\
0.23 & -0.41 & 0.25 & 0.12 \\
-1.82 & 0.45 & 1.98 & -1.47 \\
0.93 & -0.90 & -1.27 & 2.91
\end{bmatrix}
$$

The difference of log odds ratios for those cells that are at the same distance away from the diagonal is with the same value.

$$\hat{\Phi}_{12} - \hat{\Phi}_{21} = 0.13 - (-0.03) = 0.20 = \hat{\Phi}_{23} - \hat{\Phi}_{32} = \hat{\Phi}_{34} - \hat{\Phi}_{43}$$
$$\hat{\Phi}_{13} - \hat{\Phi}_{31} = \hat{\Phi}_{24} - \hat{\Phi}_{42} = -1.02$$
$$\hat{\Phi}_{14} - \hat{\Phi}_{41} = 0.20$$

Similarly, the ratio of odds ratios displays a constant term for those cells that are at the same distance away from the diagonal.

$$\frac{\hat{\theta}_{21}}{\hat{\theta}_{12}} = 1.2228 = \frac{\hat{\theta}_{32}}{\hat{\theta}_{23}} = \frac{\hat{\theta}_{43}}{\hat{\theta}_{34}}$$

$$\frac{\hat{\theta}_{31}}{\hat{\theta}_{13}} = 0.3600 = \frac{\hat{\theta}_{42}}{\hat{\theta}_{24}}$$

$$\frac{\hat{\theta}_{41}}{\hat{\theta}_{14}} = 1.2239$$

The following R codes generate $\hat{\Phi}_{ij} - \hat{\Phi}_{ji}$ of QDPS.

```
# First Position Away Diagonal
LocalOdds21_12 <- log(LocalOdds[2,1]) - log(LocalOdds[1,2])
LocalOdds32_23 <- log(LocalOdds[3,2]) - log(LocalOdds[2,3])
LocalOdds43_34 <- log(LocalOdds[4,3]) - log(LocalOdds[3,4])
LocalOdds21_12
LocalOdds32_23
LocalOdds43_34
# Second Position Away Diagonal
LocalOdds31_13 <- log(LocalOdds[3,1]) - log(LocalOdds[1,3])
LocalOdds42_24 <- log(LocalOdds[4,2]) - log(LocalOdds[2,4])
LocalOdds31_13
LocalOdds42_24
# Third Position Away Diagonal
LocalOdds41_14 <- log(LocalOdds[4,1]) - log(LocalOdds[1,4])
LocalOdds41_14
```

```
> LocalOdds21_12
[1] 0.2011555
> LocalOdds32_23            > LocalOdds31_13
[1] 0.2011555               [1] -1.021726
> LocalOdds43_34            > LocalOdds42_24         > LocalOdds41_14
[1] 0.2011555               [1] -1.021726            [1] 0.2020381
```

4.7.1 Relationships of Complete Symmetry, Conditional Symmetry, Quasi Symmetry, Diagonal Parameters Symmetry, and Quasi Diagonal Parameters Symmetry Models

While δ_k is the odds of observation falls in cell (i,j) instead of cell (j,i), as defined in DPS, a new parameter ϕ_j is added and it becomes QDPS. The relationships of S, CS, QS, DPS, and QDPS are depicted in Fig. 4.6. When $\phi_j = 1$ and $\delta_k = 1$, QDPS reduces to complete symmetry model (S). When $\phi_j = \gamma$ and $\delta_k = 1$, it becomes the conditional symmetry (CS) model. When $\phi_j = 1$, QDPS reduces to DPS. When $\delta_k = 1$, it reduces to QS. When $\delta_k = 1$, QDPS reduces to QS (Fig. 4.13).

Fig. 4.13 Relationships of S, CS, QS, DPS, and QDPS

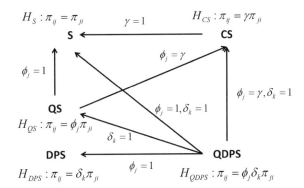

4.8   The 2-Ratios Parameters Symmetry Model

The 2-ratios parameters symmetry (2RPS) model (Tomizawa 1987) is defined as follows:

$$H_{2\text{RPS}} : \pi_{ij} = \gamma \delta^{j-i} \pi_{ji} \quad \text{for } i < j$$

where π_{ij} denote the probability that an observation falls in the cell in row i and column j, and δ^{j-i} is defined previously in DPS. As the name implies, there are two ratios that describe the main characteristics of this model. The two symbolic tables below describe the two main characteristics of 2RPS. The estimated odds for those cells that are the same distance from the diagonal are with the same value. δ describes the relationship of cells that are one position away from the diagonal, δ^2 describes the association of cells that are two positions away from the diagonal, δ^3 is three positions away, and so on. For instance, the odds of cell (1,2) to cell (2,1), cell (2,3) to cell (3,2), up to cell (5,6) to cell (6,5) have the same odds δ. More importantly, a linear relationship of log odds exists for the diagonal cells. $\ln(\delta^2) - \ln(\delta) = \ln(\delta^3) - \ln(\delta^2) = \ln(\delta^4) - \ln(\delta^3) = \ln(\delta^5) - \ln(\delta^4)$. The second characteristic is that the odds ratios in the odds ratio table enclosed by the red triangle are symmetrical, and those cells that are one location away from diagonal are a constant odds of γ (Fig. 4.14, right-hand side symbolic table). For instance, $\theta_{12} = \gamma\theta_{21}$ and $\theta_{23} = \gamma\theta_{32}$. The ratios δ^{j-i} and γ describe the two main features of 2-ratios parameters symmetry model (Fig. 4.14).

The nonstandard log-linear specification for 2RPS model is stated below. It is based on $(a^2 - a - 4)/2$ degrees of freedom.

$$\ln(\hat{m}_{ij}) = \mu + \lambda_{ij}^{CS} + \lambda_{ij}^{S} + \lambda_{ij}^{F}$$

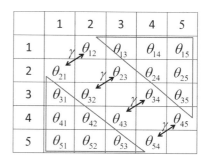

Fig. 4.14 2-ratios parameters symmetry model

$$F_{ij} = \begin{cases} |i-j|+1 & i<j \\ 1 & \text{otherwise} \end{cases}$$

For a 6×6 table, the elements of CS, S, and F are as follows:

$$\text{CS} = \begin{bmatrix} 1 & 1 & 1 & 1 & 1 & 1 \\ 2 & 1 & 1 & 1 & 1 & 1 \\ 2 & 2 & 1 & 1 & 1 & 1 \\ 2 & 2 & 2 & 1 & 1 & 1 \\ 2 & 2 & 2 & 2 & 1 & 1 \\ 2 & 2 & 2 & 2 & 2 & 1 \end{bmatrix}, \quad S = \begin{bmatrix} 1 & 2 & 3 & 4 & 5 & 6 \\ 2 & 7 & 8 & 9 & 10 & 11 \\ 3 & 8 & 12 & 13 & 14 & 15 \\ 4 & 9 & 13 & 16 & 17 & 18 \\ 5 & 10 & 14 & 17 & 19 & 20 \\ 6 & 11 & 15 & 18 & 20 & 21 \end{bmatrix},$$

$$F = \begin{bmatrix} 1 & 2 & 3 & 4 & 5 & 6 \\ 1 & 1 & 2 & 3 & 4 & 5 \\ 1 & 1 & 1 & 2 & 3 & 4 \\ 1 & 1 & 1 & 1 & 2 & 3 \\ 1 & 1 & 1 & 1 & 1 & 2 \\ 1 & 1 & 1 & 1 & 1 & 1 \end{bmatrix}.$$

Table 4.11 shows the cross-tabulation of father and son occupation status.

The R syntax to generate 2-ratios parameters symmetry model for the above table is shown below.

```
FatherSonOcc <- c(33,  9,  6,  2,   2,   1,
                  18, 25, 11,  9,   4,   2,
                  16, 21, 68, 20,  12,   8,
                   6, 23, 40, 74,  35,  14,
                  10, 14, 30, 71, 132,  55,
                   2,  9, 24, 35, 110, 120)
S <- c(1, 2,  3,  4,  5,  6,
       2, 7,  8,  9, 10, 11,
       3, 8, 12, 13, 14, 15,
       4, 9, 13, 16, 17, 18,
       5, 10, 14, 17, 19, 20,
```

Table 4.11 Father and son occupation status	Son occupation status	Father occupation status					
		1	2	3	4	5	6
	1	33	9	6	2	2	1
	2	18	25	11	9	4	2
	3	16	21	68	20	12	8
	4	6	23	40	74	35	14
	5	10	14	30	71	132	55
	6	2	9	24	35	110	120

```
        6,11,15,18,20,21)
S <- as.factor(S)
F <- c(1, 2, 3, 4, 5, 6,
       1, 1, 2, 3, 4, 5,
       1, 1, 1, 2, 3, 4,
       1, 1, 1, 1, 2, 3,
       1, 1, 1, 1, 1, 2,
       1, 1, 1, 1, 1, 1)
CS <- c( 1, 1, 1, 1, 1, 1,
         2, 1, 1, 1, 1, 1,
         2, 2, 1, 1, 1, 1,
         2, 2, 2, 1, 1, 1,
         2, 2, 2, 2, 1, 1,
         2, 2, 2, 2, 2, 1)
CS <- as.factor(CS)
RPS2 <- glm(FatherSonOcc~S+F+CS,family=poisson)
summary(RPS2)
model.summary(RPS2)
```

The estimated coefficients for 2RPS model are printed below. The two estimated ratio coefficients are F (-0.2272) and CS2 (0.4694).

```
> summary(RPS2)

Call:
glm(formula = FatherSonOcc ~ S + F + CS, family = poisson)

Deviance Residuals:
     Min        1Q    Median        3Q       Max
 -0.32657  -0.03141   0.00000   0.02252   0.61840

Coefficients:
             Estimate Std. Error z value Pr(>|z|)
(Intercept)    3.7236     0.2065  18.032  < 2e-16 ***
S2            -1.0744     0.3039  -3.535 0.000407 ***
S3            -1.2093     0.3354  -3.605 0.000312 ***
S4            -2.1614     0.4493  -4.810 1.51e-06 ***
S5            -1.7059     0.4092  -4.169 3.06e-05 ***
S6            -3.0504     0.6496  -4.696 2.65e-06 ***
S7            -0.2776     0.2651  -1.047 0.295060
S8            -0.9045     0.2942  -3.074 0.002111 **
S9            -0.8346     0.3135  -2.662 0.007773 **
S10           -1.3505     0.3640  -3.711 0.000207 ***
S11           -1.7929     0.4184  -4.286 1.82e-05 ***
S12            0.7230     0.2122   3.408 0.000655 ***
S13           -0.2759     0.2683  -1.028 0.303785
S14           -0.5627     0.3014  -1.867 0.061970 .
S15           -0.7751     0.3289  -2.357 0.018429 *
S16            0.8076     0.2093   3.858 0.000114 ***
S17            0.2932     0.2545   1.152 0.249325
S18           -0.4085     0.2957  -1.381 0.167207
S19            1.3863     0.1946   7.123 1.06e-12 ***
S20            0.7357     0.2478   2.969 0.002986 **
S21            1.2910     0.1966   6.568 5.10e-11 ***
F             -0.2271     0.1111  -2.044 0.040917 *
CS2            0.4694     0.1867   2.514 0.011931 *
---
Signif. codes:  0 '***' 0.001 '**' 0.01 '*' 0.05 '.' 0.1 ' ' 1
```

The estimated values and odds ratios of 2RPS are stated below.

$$
\hat{m}_{ij} = \begin{bmatrix}
33 & 8.98 & 6.25 & 1.92 & 2.42 & 0.50 \\
18.02 & 25 & 10.64 & 9.10 & 4.33 & 2.22 \\
15.75 & 21.36 & 68 & 19.96 & 11.94 & 7.69 \\
6.08 & 22.90 & 40.04 & 74 & 35.25 & 13.93 \\
9.58 & 13.67 & 30.06 & 70.75 & 132 & 54.88 \\
2.50 & 8.78 & 24.31 & 35.07 & 110.12 & 120
\end{bmatrix}
$$

$$
\hat{\theta}_{ij} = \begin{bmatrix}
5.10 & 0.61 & 2.78 & 0.38 & 2.46 \\
0.98 & 7.48 & 0.34 & 1.26 & 1.26 \\
2.78 & 0.55 & 6.30 & 0.80 & 0.61 \\
0.38 & 1.26 & 1.27 & 3.92 & 1.05 \\
2.46 & 1.26 & 0.61 & 1.68 & 2.62
\end{bmatrix}
$$

The estimated coefficient of F that gives the relationship of cells from the diagonal position is 0.2271 in log odds differences. The following shows the estimates of δ, δ^2–δ^5.

$$
\delta = \frac{18.02}{8.98} = \frac{21.36}{10.64} = \frac{40.04}{19.96} = \frac{70.75}{35.25} = \frac{110.12}{54.88} = 2.01
$$

$$
\delta^2 = \frac{15.75}{6.25} = \frac{22.90}{9.10} = \frac{30.06}{11.94} = \frac{35.07}{13.93} = 2.52
$$

$$
\delta^3 = \frac{6.08}{1.92} = \frac{4.33}{11.94} = \frac{7.69}{24.31} = 3.16
$$

$$
\delta^4 = \frac{9.58}{2.42} = \frac{8.78}{2.22} = 3.97
$$

$$
\delta^5 = \frac{2.50}{0.50} = 4.981
$$

The estimates of $\ln(\delta)$, $\ln(\delta^2)$ to $\ln(\delta^5)$ are given below.

$$
\ln(\delta) = 0.70; \quad \ln(\delta^2) = 0.92; \quad \ln(\delta^3) = 1.15;
$$
$$
\ln(\delta^4) = 1.37; \quad \ln(\delta^5) = 1.60
$$

The difference of $\ln(\delta^2)$ and $\ln(\delta)$ is equal to the estimated coefficient of F (−0.2271). Similarly, this association applies to the difference of $\ln(\delta^3)$ and $\ln(\delta^2)$, the difference of $\ln(\delta^4)$ and $\ln(\delta^3)$, and the difference of $\ln(\delta^5)$ and $\ln(\delta^4)$.

$$
\ln(\delta^2) - \ln(\delta) = \ln(\delta^3) - \ln(\delta^2)
$$
$$
= \ln(\delta^4) - \ln(\delta^3) = \ln(\delta^5) - \ln(\delta^4) = 0.2271
$$

The exponential of the estimated value of CS (0.4694), γ, is the ratio of odds ratios for those cells that are one position away from the diagonal, as shown below.

$$\frac{\hat{\theta}_{21}}{\hat{\theta}_{12}} = \frac{0.98}{0.61} = \frac{\hat{\theta}_{32}}{\hat{\theta}_{23}} = \frac{0.55}{0.34} = \frac{\hat{\theta}_{43}}{\hat{\theta}_{34}} = \frac{1.27}{0.80} = \frac{\hat{\theta}_{54}}{\hat{\theta}_{45}} = \frac{1.68}{1.05} = \exp(0.4694) = 1.60$$

The MLE of $\hat{m}_{ij}$ and $\hat{m}_{ji}$ satisfies the following specification that the estimated value of $\hat{m}_{ij} + \hat{m}_{ji}$ is equal to the raw data $n_{ij} + n_{ji}$ (Lawal 2001). For instance, $\hat{m}_{12} + \hat{m}_{21} = 8.98 + 18.02 = n_{12} + n_{21} = 18 + 9 = 27$,   and $\hat{m}_{46} + \hat{m}_{64} = 13.93 + 35.07 = n_{46} + n_{64} = 14 + 35 = 49$.

$$\hat{m}_{ij} + \hat{m}_{ji} = n_{ij} + n_{ji}$$

$$\sum_{i=1}^{I}\sum_{j=1}^{I} j\hat{m}_{ij} = \sum_{i=1}^{I}\sum_{j=1}^{I} jn_{ij} \quad \text{and}$$

$$\sum_{i<j} \hat{m}_{ij}\{I - 2(j-i)\} = \sum_{i<j} n_{ij}\{I - 2(j-i)\} \quad \text{for } 1 \leq (i,j) \leq I$$

4.8.1   Relationships of Conditional Symmetry, Linear Diagonal Parameters Symmetry, and 2-Ratios Parameters Symmetry Models

The relationships of conditional symmetry (CS), linear diagonals parameter symmetry (LDPS), and 2-ratios parameters symmetry (2RPS) models are depicted in Fig. 4.15. Both CS and LDPS are special cases of 2RPS model; i.e., 2RPS is an extension of the LDPS and CS models. When $\delta = 1$, 2RPS reduces to CS, and when $\gamma = 1$, 2RPS reduces to LDPS. The log odds for both 2RPS and LDPS are proportional to the distance from the main diagonal. When $\log \delta = 0$, 2RPS becomes CS.

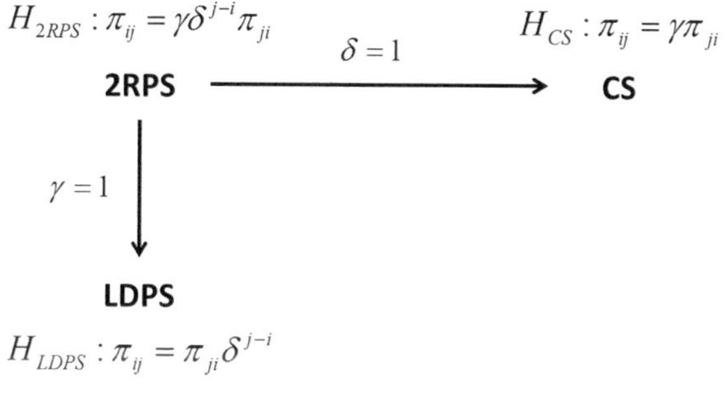

Fig. 4.15 Relationships of conditional symmetry, linear parameters symmetry, and 2-ratios parameters symmetry models

$$H_{2RPS}: \pi_{ij} = \gamma\delta^{j-i}\pi_{ji}$$
$$\xrightarrow{\delta=1}$$
2RPS   **CS** $H_{CS}: \pi_{ij} = \gamma\pi_{ji}$

$\gamma = 1$ ↓

LDPS

$$H_{LDPS}: \pi_{ij} = \pi_{ji}\delta^{j-i}$$

4.9 Quasi Conditional Symmetry Model

The quasi conditional symmetry model (QCS) is also known as extended quasi symmetry model (Tomizawa 1987) and uniform skew symmetric level model (Yamagushi 1990). It has the following formulation with degrees of freedom $a(a-3)/2$.

$$H_{QCS} : \pi_{ij} = \gamma \alpha_j \pi_{ji} \quad \text{for } i<j$$

where γ is an unspecified parameter.

The characteristics of quasi conditional symmetry model is that

$$\pi_{ij}\pi_{jk}\pi_{ki} = \gamma \pi_{ji}\pi_{kj}\pi_{ik} \quad i<j<k$$

$$\Omega_{(ij,jk)} = \Omega_{(ij,ik)} = \gamma \quad 1 \le i<j<k \le I$$

$$\text{where } \Omega_{(ij,st)} = \frac{\Theta_{(ij,st)}}{\Theta_{(st,ij)}}, \quad \Theta_{(ij,st)} = \frac{\pi_{is}\pi_{jt}}{\pi_{js}\pi_{it}}$$

The estimate $\Omega_{(ij,jk)}$ is defined as follows:

$$\hat{\Omega}_{(ij,st)} = \frac{\hat{m}_{is} \times \hat{m}_{jt} \times \hat{m}_{ti} \times \hat{m}_{sj}}{\hat{m}_{si} \times \hat{m}_{tj} \times \hat{m}_{it} \times \hat{m}_{js}}$$

The two symbolic tables for quasi conditional symmetry model describe the above-mentioned characteristics of this model. The right-hand side symbolic table of Fig. 4.16 states that those diagonal odds ratios $(\theta_{11}, \theta_{22}, \theta_{33}, \theta_{44}, \theta_{55})$ of this symbolic table have their separate estimated values—those cells left blank. Those cells one position off the diagonal cells are with a constant term γ to their diagonally symmetrical cells, that is $\theta_{12} = \gamma\theta_{21}, \ldots, \theta_{45} = \gamma\theta_{54}$. The rest of the odds ratios are in symmetrical, referring to those cells two positions away from the diagonal.

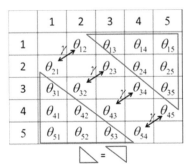

Fig. 4.16 Quasi conditional symmetry model

For instance, $\theta_{13} = \theta_{31}$ and $\theta_{14} = \theta_{41}$. The left-hand side of Fig. 4.15 shows the cells that satisfy the condition: $\Omega_{(ij,jk)} = \Omega_{(ij,ik)} = \gamma$. The multiplication of those cells in triangle over the multiplication of those cells in circle at the upper left is a constant value of γ. Similarly, the multiplication of those cells in triangle over the multiplication of those cells in circle at the lower right also produces the constant value γ.

The nonstandard log-linear specification of conditional symmetry model, with $a(a-3)/2$ degrees of freedom, is specified as follows:

$$\ln(\hat{m}_{ij}) = \mu + \lambda_i^R + \lambda_j^C + \lambda_{ij}^S + \eta^{CS}$$

The factor variables S, R, and C and regression variables CS are specified as follows:

$$R = \begin{bmatrix} 1 & 2 & 3 & 4 & 5 & 6 & 7 & 8 \\ 1 & 2 & 3 & 4 & 5 & 6 & 7 & 8 \\ 1 & 2 & 3 & 4 & 5 & 6 & 7 & 8 \\ 1 & 2 & 3 & 4 & 5 & 6 & 7 & 8 \\ 1 & 2 & 3 & 4 & 5 & 6 & 7 & 8 \\ 1 & 2 & 3 & 4 & 5 & 6 & 7 & 8 \\ 1 & 2 & 3 & 4 & 5 & 6 & 7 & 8 \\ 1 & 2 & 3 & 4 & 5 & 6 & 7 & 8 \end{bmatrix}, \quad C = \begin{bmatrix} 1 & 1 & 1 & 1 & 1 & 1 & 1 & 1 \\ 2 & 2 & 2 & 2 & 2 & 2 & 2 & 2 \\ 3 & 3 & 3 & 3 & 3 & 3 & 3 & 3 \\ 4 & 4 & 4 & 4 & 4 & 4 & 4 & 4 \\ 5 & 5 & 5 & 5 & 5 & 5 & 5 & 5 \\ 6 & 6 & 6 & 6 & 6 & 6 & 6 & 6 \\ 7 & 7 & 7 & 7 & 7 & 7 & 7 & 7 \\ 8 & 8 & 8 & 8 & 8 & 8 & 8 & 8 \end{bmatrix},$$

$$S = \begin{bmatrix} 1 & 2 & 3 & 4 & 5 & 6 & 7 & 8 \\ 2 & 9 & 10 & 11 & 12 & 13 & 14 & 15 \\ 3 & 10 & 16 & 17 & 18 & 19 & 20 & 21 \\ 4 & 11 & 17 & 22 & 23 & 24 & 25 & 26 \\ 5 & 12 & 18 & 23 & 27 & 28 & 29 & 30 \\ 6 & 13 & 19 & 24 & 28 & 31 & 32 & 33 \\ 7 & 14 & 20 & 25 & 29 & 32 & 34 & 35 \\ 8 & 15 & 21 & 26 & 30 & 33 & 35 & 36 \end{bmatrix}, \quad CS = \begin{bmatrix} 1 & 1 & 1 & 1 & 1 & 1 & 1 & 1 \\ 2 & 1 & 1 & 1 & 1 & 1 & 1 & 1 \\ 2 & 2 & 1 & 1 & 1 & 1 & 1 & 1 \\ 2 & 2 & 2 & 1 & 1 & 1 & 1 & 1 \\ 2 & 2 & 2 & 2 & 1 & 1 & 1 & 1 \\ 2 & 2 & 2 & 2 & 2 & 1 & 1 & 1 \\ 2 & 2 & 2 & 2 & 2 & 2 & 1 & 1 \\ 2 & 2 & 2 & 2 & 2 & 2 & 2 & 1 \end{bmatrix}$$

Table 4.12 tabulates the scores of two psychologists 1 and 2 after assessing 4676 patients for score of 1 being the lowest and 8 the highest in depression.

Table 4.12 Score by psychologists 1 and 2

Score by psychiatrist 2	Score by psychiatrist 1							
	1	2	3	4	5	6	7	8
1	372	30	10	11	20	20	14	20
2	1	650	10	5	4	3	1	1
3	1	5	840	22	3	5	3	5
4	1	4	4	1075	60	11	2	8
5	0	6	1	8	398	60	11	10
6	1	5	4	5	10	339	45	28
7	0	3	2	3	2	7	78	35
8	3	20	16	15	10	15	15	300

The R syntax below generates quasi conditional symmetry model for the above table.

```
Score <- c(372,  30,   10,    11,  20,  20,  14,  20,
             1, 650,   10,     5,   4,   3,   1,   1,
             1,   5,  840,    22,   3,   5,   3,   5,
             1,   4,    4,  1075,  60,  11,   2,   8,
             0,   6,    1,     8, 398,  60,  11,  10,
             1,   5,    4,     5,  10, 339,  45,  28,
             0,   3,    2,     3,   2,   7,  78,  35,
             3,  20,   16,    15,  10,  15,  15, 300)
r<-gl(8,8)
c<-gl(8,1,length=64)
s <- c(1, 2, 3, 4, 5, 6, 7, 8,
       2,  9, 10, 11, 12, 13, 14, 15,
       3, 10, 16, 17, 18, 19, 20, 21,
       4, 11, 17, 22, 23, 24, 25, 26,
       5, 12, 18, 23, 27, 28, 29, 30,
       6, 13, 19, 24, 28, 31, 32, 33,
       7, 14, 20, 25, 29, 32, 34, 35,
       8, 15, 21, 26, 30, 33, 35, 36)
s <- as.factor(s)
cs <- c(1, 1, 1, 1, 1, 1, 1, 1,
        2, 1, 1, 1, 1, 1, 1, 1,
        2, 2, 1, 1, 1, 1, 1, 1,
        2, 2, 2, 1, 1, 1, 1, 1,
        2, 2, 2, 2, 1, 1, 1, 1,
        2, 2, 2, 2, 2, 1, 1, 1,
        2, 2, 2, 2, 2, 2, 1, 1,
        2, 2, 2, 2, 2, 2, 2, 1)
QCS<-glm(Score~r+c+s+cs,family=poisson)
summary(QCS)
model.summary(QCS)
```

The estimated coefficients of quasi conditional symmetry model are printed below.

```
> summary(QCS)

Call:
glm(formula = Score ~ r + c + s + cs, family = poisson)

Deviance Residuals:
    Min       1Q    Median       3Q      Max
 -1.2781  -0.1771    0.0000   0.1283   1.8448

Coefficients: (7 not defined because of singularities)
            Estimate Std. Error z value Pr(>|z|)
(Intercept)  8.10276    0.20557  39.416  < 2e-16 ***
r2          -4.91225    0.41958 -11.707  < 2e-16 ***
r3          -3.72737    0.36668 -10.165  < 2e-16 ***
r4          -3.28128    0.33394  -9.826  < 2e-16 ***
r5          -3.14808    0.34318  -9.173  < 2e-16 ***
r6          -2.13203    0.30421  -7.008 2.41e-12 ***
r7          -1.88726    0.30376  -6.213 5.20e-10 ***
r8           0.25990    0.24917   1.043 0.296919
c2          -1.03491    0.31301  -3.306 0.000945 ***
c3          -1.25830    0.32210  -3.907 9.36e-05 ***
c4          -1.33423    0.31938  -4.178 2.95e-05 ***
c5          -1.68561    0.32715  -5.152 2.57e-07 ***
c6          -1.24996    0.28038  -4.458 8.27e-06 ***
c7          -1.28986    0.28648  -4.502 6.72e-06 ***
c8          -0.47501    0.24917  -1.906 0.056600 .
s2          -1.45233    0.36014  -4.033 5.51e-05 ***
s3          -2.27219    0.43860  -5.181 2.21e-07 ***
s4          -2.11570    0.42613  -4.965 6.87e-07 ***
s5          -1.26330    0.38925  -3.246 0.001172 **
s6          -1.66997    0.34284  -4.871 1.11e-06 ***
s7          -2.05009    0.37623  -5.449 5.07e-08 ***
s8          -2.51931    0.27567  -9.139  < 2e-16 ***
s9           6.50524    0.52100  12.486  < 2e-16 ***
s10          2.58097    0.45473   5.676 1.38e-08 ***
s11          1.95043    0.48203   4.046 5.20e-05 ***
s12          2.16624    0.46531   4.656 3.23e-06 ***
s13          1.14377    0.45351   2.522 0.011668 *
s14          0.28512    0.56992   0.500 0.616876
s15               NA         NA      NA       NA
s16          5.80017    0.48584  11.938  < 2e-16 ***
s17          2.22703    0.41631   5.349 8.82e-08 ***
s18          0.61179    0.61768   0.990 0.321953
s19          0.81706    0.45012   1.815 0.069494 .
s20          0.15859    0.53281   0.298 0.765973
s21               NA         NA      NA       NA
s22          5.67669    0.46186  12.291  < 2e-16 ***
s23          3.09961    0.37383   8.291  < 2e-16 ***
s24          1.10229    0.38911   2.833 0.004613 **
s25         -0.09893    0.53208  -0.186 0.852499
s26               NA         NA      NA       NA
s27          4.90125    0.49535   9.895  < 2e-16 ***
s28          2.54432    0.34235   7.432 1.07e-13 ***
s29          0.84698    0.41704   2.031 0.042260 *
s30               NA         NA      NA       NA
s31          3.28910    0.36953   8.901  < 2e-16 ***
s32          1.31475    0.31253   4.207 2.59e-05 ***
s33               NA         NA      NA       NA
s34          1.61494    0.36068   4.478 7.55e-06 ***
s35               NA         NA      NA       NA
s36               NA         NA      NA       NA
cs          -2.18386    0.19893 -10.978  < 2e-16 ***
---
Signif. codes:  0 '***' 0.001 '**' 0.01 '*' 0.05 '.' 0.1 ' ' 1
```

The fitted values and odds ratio are listed below.

$$
\hat{m}_{ij} =
\begin{bmatrix}
372 & 30.98 & 10.90 & 11.81 & 19.49 & 20.06 & 13.18 & 18.63 \\
0.07 & 650 & 10.27 & 5.07 & 4.43 & 2.46 & 1.00 & 1.70 \\
0.10 & 4.73 & 840 & 21.85 & 3.06 & 5.80 & 2.89 & 5.57 \\
0.19 & 3.93 & 4.15 & 1075 & 57.49 & 12.06 & 3.49 & 8.69 \\
0.51 & 5.57 & 0.94 & 10.51 & 398 & 58.28 & 10.26 & 9.93 \\
0.94 & 5.54 & 3.20 & 3.94 & 11.72 & 339 & 45.23 & 27.44 \\
0.82 & 3.00 & 2.11 & 1.51 & 2.74 & 6.77 & 78 & 35.04 \\
4.37 & 19.30 & 15.43 & 14.31 & 10.07 & 15.56 & 14.96 & 300
\end{bmatrix}
$$

$$\hat{\theta}_{ij} = \begin{bmatrix} 108418 & 0.0449 & 0.4552 & 0.5291 & 0.5402 & 0.6197 & 1.2021 \\ 0.0051 & 11241 & 0.0527 & 0.1602 & 3.4135 & 1.2216 & 1.1349 \\ 0.4552 & 0.0059 & 9963 & 0.3822 & 0.1105 & 0.5812 & 1.2937 \\ 0.5291 & 0.1602 & 0.0430 & 708 & 0.6980 & 0.6089 & 0.3883 \\ 0.5402 & 3.4135 & 0.1105 & 0.0786 & 198 & 0.7580 & 0.6264 \\ 0.6197 & 1.2216 & 0.5812 & 0.6089 & 0.0854 & 86 & 0.7407 \\ 1.2021 & 1.1349 & 1.2937 & 0.3883 & 0.6264 & 0.0834 & 45 \end{bmatrix}$$

The estimated values of γ can derived using the formula $\gamma = \frac{\pi_{ij}\pi_{jk}\pi_{ki}}{\pi_{ji}\pi_{kj}\pi_{ik}}$. Alternatively, the value of γ can be extracted from the estimated coefficient CS after taking the exponential, $\exp(2.18386) = 8.88$. It can also be estimated using those odds ratios that are one position away from the diagonal, as shown below.

$$\gamma = \frac{\hat{m}_{12}\hat{m}_{23}\hat{m}_{31}}{\hat{m}_{21}\hat{m}_{32}\hat{m}_{13}} = \frac{30.98 \times 10.27 \times 0.10}{0.07 \times 4.73 \times 10.90} = \frac{\hat{m}_{13}\hat{m}_{34}\hat{m}_{41}}{\hat{m}_{31}\hat{m}_{43}\hat{m}_{14}}$$
$$= \frac{10.90 \times 21.85 \times 0.19}{0.10 \times 4.15 \times 11.81} = \cdots = \frac{\hat{m}_{45}\hat{m}_{56}\hat{m}_{64}}{\hat{m}_{54}\hat{m}_{65}\hat{m}_{46}}$$
$$= \frac{57.49 \times 58.28 \times 3.04}{10.51 \times 11.73 \times 12.06} = \cdots = 8.88$$
$$\gamma = \frac{\hat{\theta}_{12}}{\hat{\theta}_{21}} = \frac{0.049}{0.0051} = \frac{\hat{\theta}_{23}}{\hat{\theta}_{32}} = \frac{0.0527}{0.0059} = \cdots = \frac{\hat{\theta}_{67}}{\hat{\theta}_{76}} = \frac{0.7407}{0.0834} = 8.88$$

4.10 Quasi Odds Symmetry Model

Quasi odds symmetry model is the generalization of odds symmetry model (Tomizawa 1985). Bishop et al. (1975) refer it as adjusted quasi symmetry model, and Yamagushi (1990) refers it as middle value effect skew symmetry model. The quasi odds symmetry model has the following formulation with degrees of freedom $(a - 2)(a - 3)/2$,

$$H_{QOS} : \pi_{ij}\pi_{jk}\pi_{ki} = \gamma_j \pi_{ji}\pi_{kj}\pi_{ik} \quad 1 \le i < j < k \le a$$

where γ_j are unspecified parameters.

The two symbolic tables below show the main characteristics of quasi odds symmetry model. The left symbolic table specifies the cell probabilities to produce the parameters γ_j. γ_j are the multiplication of the cells enclosed in triangle divided by the cells enclosed in circle. For instance, the green colored three cells of triangles divided by the three green cells of circles derive the parameter γ_1, and the red enclosed set of triangles divided by the set of circles derive the parameter γ_4. The second symbolic table on the right shows the characteristic of quasi odds symmetry model in terms of odds ratios. The ratio of those odds ratios that is one position off the diagonal is γ_j, depending on the position of the cell at column j. That is, $\theta_{12} = \gamma_1\theta_{21}$, $\theta_{23} = \gamma_2\theta_{32}$, $\theta_{34} = \gamma_3\theta_{43}$, and $\theta_{45} = \gamma_4\theta_{54}$. Those cells enclosed with the two red triangles have the same odds ratios for cells in the same opposite diagonal position. For instance, $\theta_{13} = \theta_{31}$ and $\theta_{15} = \theta_{51}$ (Fig. 4.17).

Quasi odds symmetry model has two forms. Odd symmetry model I and odds symmetry model II are the two cases of quasi odds symmetry model. The non-standard log-linear models for the two quasi odds symmetry models are stated below.

$$\ln(\hat{m}_{ij}) = \mu + \lambda_i^R + \lambda_j^C + \lambda_{ij}^S + \lambda_{ij}^{OS1} \quad \text{for } QS + OS1$$
$$\ln(\hat{m}_{ij}) = \mu + \lambda_i^R + \lambda_j^C + \lambda_{ij}^S + \lambda_{ij}^{OS2} \quad \text{for } QS + OS2$$

Table 4.13 tabulates the 923 assessment ratings of raters A and B.

The R syntax to generate the two quasi odds symmetry models for the above table is stated below.

```
QOS1Data <- c(56, 10, 12, 20, 23, 11,
              21, 20, 44, 22, 27, 40,
              40, 21, 70, 20, 13, 16,
              15, 12, 21, 40, 22, 13,
              40, 13, 12, 34, 55, 6,
              21, 22, 25, 20, 10, 56)
r <- gl(6,6)
c <- gl(6,1,length=36)
s <- c(1, 2, 3, 4, 5, 6,
```

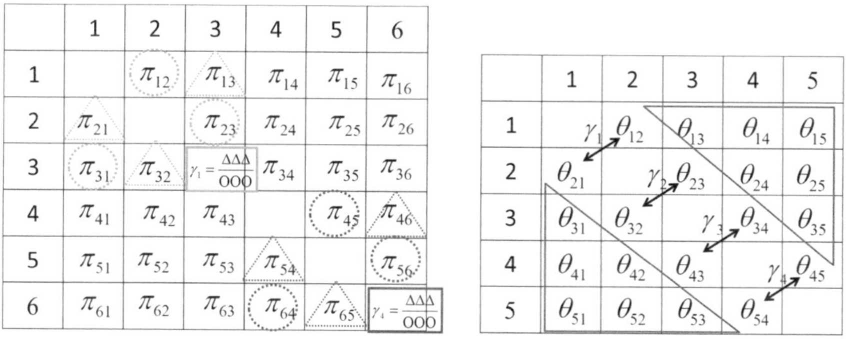

	1	2	3	4	5	6
1		π_{12}	π_{13}	π_{14}	π_{15}	π_{16}
2	π_{21}		π_{23}	π_{24}	π_{25}	π_{26}
3	π_{31}	π_{32}	$\gamma_i=\frac{\triangle\triangle\triangle}{OOO}$	π_{34}	π_{35}	π_{36}
4	π_{41}	π_{42}	π_{43}		π_{45}	π_{46}
5	π_{51}	π_{52}	π_{53}	π_{54}		π_{56}
6	π_{61}	π_{62}	π_{63}	π_{64}	π_{65}	$\gamma_i=\frac{\triangle\triangle\triangle}{OOO}$

	1	2	3	4	5
1	γ_1	θ_{12}	θ_{13}	θ_{14}	θ_{15}
2	θ_{21}	γ_2	θ_{23}	θ_{24}	θ_{25}
3	θ_{31}	θ_{32}	γ_3	θ_{34}	θ_{35}
4	θ_{41}	θ_{42}	θ_{43}	γ_4	θ_{45}
5	θ_{51}	θ_{52}	θ_{53}	θ_{54}	

Fig. 4.17 Quasi odds symmetry model

Table 4.13 Rater A × rater B

Rater B	Rater A					
	1	2	3	4	5	6
1	56	10	12	20	23	11
2	21	20	44	22	27	40
3	40	21	70	20	13	16
4	15	12	21	40	22	13
5	40	13	12	34	55	6
6	21	22	25	20	10	56

```
2,  7,  8,  9,10,11,
3,  8,12,13,14,15,
4,  9,13,16,17,18,
5,10,14,17,19,20,
6,11,15,18,20,21)
s <- as.factor(s)
os1<- c(11, 1, 1, 1, 1, 1,
         6,11, 2, 2, 2, 2,
         6, 7,11, 3, 3, 3,
         6, 7, 8,11, 4, 4,
         6, 7, 8, 9,11, 5,
         6, 7, 8, 9,10,11)
os1 <- as.factor(os1)
QOS1 <-glm(QOS1Data~r+c+s+os1,family=poisson)
summary(QOS1)
model.summary(QOS1)
```

The estimated coefficients for the quasi odds symmetry model are stated below.

```
> summary(QOS1)

Call:
glm(formula = QOS1Data ~ r + c + s + os1, family = poisson)

Deviance Residuals:
    Min       1Q   Median       3Q      Max
-1.0523  -0.1419   0.0000   0.1192   1.0810

Coefficients: (11 not defined because of singularities)
            Estimate Std. Error z value Pr(>|z|)
(Intercept)  4.02535    0.13363  30.123  < 2e-16 ***
r2          -0.53984    0.43430  -1.243 0.213871
r3          -0.35379    0.29514  -1.199 0.230644
r4          -1.10018    0.30869  -3.564 0.000365 ***
r5          -1.54770    0.37768  -4.098 4.17e-05 ***
r6           0.33424    0.14631   2.284 0.022345 *
c2          -1.28177    0.20249  -6.330 2.45e-10 ***
c3          -1.16810    0.21765  -5.367 8.01e-08 ***
c4          -1.32430    0.23213  -5.705 1.16e-08 ***
c5          -2.05701    0.34843  -5.904 3.56e-09 ***
c6          -0.33424    0.14631  -2.284 0.022345 *
s2          -0.44099    0.37550  -1.174 0.240231
s3          -0.08701    0.28747  -0.303 0.762145
s4           0.04282    0.30167   0.142 0.887119
s5           1.19491    0.37543   3.183 0.001459 **
s6          -1.30761    0.20365  -6.421 1.36e-10 ***
s7           0.79199    0.58632   1.351 0.176763
s8           0.82759    0.29661   2.790 0.005268 **
s9           0.52354    0.32415   1.615 0.106291
s10          1.33742    0.40576   3.296 0.000980 ***
s11               NA         NA      NA       NA
s12          1.74503    0.45968   3.796 0.000147 ***
s13          1.22940    0.34603   3.553 0.000381 ***
s14          1.31942    0.44619   2.957 0.003105 **
s15               NA         NA      NA       NA
s16          2.08801    0.49690   4.202 2.64e-05 ***
s17          2.34898    0.43088   5.452 4.99e-08 ***
s18               NA         NA      NA       NA
s19          3.58669    0.69821   5.137 2.79e-07 ***
s20               NA         NA      NA       NA
s21               NA         NA      NA       NA
os12         0.54480    0.44236   1.232 0.218104
os13        -0.52340    0.30722  -1.704 0.088443 .
os14        -0.09008    0.32478  -0.277 0.781516
os15        -0.35165    0.56402  -0.623 0.532972
os16              NA         NA      NA       NA
os17              NA         NA      NA       NA
os18              NA         NA      NA       NA
os19              NA         NA      NA       NA
os110             NA         NA      NA       NA
os111             NA         NA      NA       NA
---
Signif. codes:  0 '***' 0.001 '**' 0.01 '*' 0.05 '.' 0.1 ' ' 1
```

The estimated values and odds ratios are printed below.

$$\hat{m}_{ij} = \begin{bmatrix} 56 & 10 & 15.96 & 15.55 & 23.65 & 10.84 \\ 21 & 20 & 40.04 & 25.27 & 27.41 & 40.29 \\ 36.04 & 24.96 & 70 & 21.18 & 11.14 & 16.68 \\ 19.45 & 8.73 & 19.82 & 40 & 22.81 & 12.19 \\ 39.35 & 12.59 & 13.86 & 33.19 & 55 & 6 \\ 21.16 & 21.71 & 24.32 & 20.81 & 10 & 56 \end{bmatrix}$$

$$\hat{\theta}_{ij} = \begin{bmatrix} 5.33 & 1.25 & 0.65 & 0.71 & 3.21 \\ 0.73 & 1.40 & 0.48 & 0.48 & 1.02 \\ 0.65 & 0.81 & 6.67 & 1.08 & 0.36 \\ 0.71 & 0.48 & 1.19 & 2.91 & 0.20 \\ 3.21 & 1.02 & 0.36 & 0.29 & 51.33 \end{bmatrix}$$

There are three ways to estimate the value of γ_j. Table 4.14 shows the working to derive γ_j.

(1) From the model coefficient estimates,
(2) From estimated frequencies,
(3) From estimated odds ratio.

Another property of QOS is that it preserves the marginal totals of both the lower left and the upper right of observed frequency. Also, the marginal row total of raw frequencies is the same as the estimated row total. This applies to the column total as well (Tomizawa 1985). For instance,

$$\hat{m}_{1.} = 10 + 15.96 + 15.54 + 23.64 + 10.84$$
$$= n_{1.} = 10 + 12 + 20 + 23 + 11 = 76,$$

$$\hat{m}_{46} + \hat{m}_{64} = 12.19 + 20.81$$
$$= n_{46} + n_{64} = 13 + 10 = 33.$$

Table 4.14 Estimated coefficients, frequencies, and odds ratios

γ_j	Factor	Estimated coefficient (1)	Estimated frequency (2)	Estimated odds ratios (3)
γ_1	os12	Exp(0.5448) = 1.72	$\frac{\hat{m}_{12}\hat{m}_{23}\hat{m}_{31}}{\hat{m}_{21}\hat{m}_{32}\hat{m}_{13}} = \frac{10\times40\times36}{21\times25\times16} = 1.72$	$\frac{\hat{\theta}_{12}}{\hat{\theta}_{21}} = \frac{1.2541}{0.7273} = 1.72$
γ_2	os13	Exp(−0.5234) = 0.59	$\frac{\hat{m}_{23}\hat{m}_{34}\hat{m}_{42}}{\hat{m}_{32}\hat{m}_{43}\hat{m}_{24}} = \frac{40\times21\times9}{25\times20\times25} = 0.59$	$\frac{\hat{\theta}_{23}}{\hat{\theta}_{32}} = \frac{0.4795}{0.8093} = 0.59$
γ_3	os14	Exp(−0.09008) = 0.92	$\frac{\hat{m}_{34}\hat{m}_{45}\hat{m}_{53}}{\hat{m}_{43}\hat{m}_{54}\hat{m}_{35}} = \frac{21\times23\times14}{20\times33\times11} = 0.92$	$\frac{\hat{\theta}_{34}}{\hat{\theta}_{43}} = \frac{1.0842}{1.1864} = 0.92$
γ_4	os15	Exp(−0.35165) = 0.70	$\frac{\hat{m}_{45}\hat{m}_{56}\hat{m}_{64}}{\hat{m}_{54}\hat{m}_{65}\hat{m}_{46}} = \frac{23\times6\times21}{33\times10\times12} = 0.70$	$\frac{\hat{\theta}_{45}}{\hat{\theta}_{54}} = \frac{0.2041}{0.2900} = 0.70$

$$\hat{m}_{i.} = n_{i.} \quad 1 \le i \le a$$
$$\hat{m}_{.j} = n_{.j} \quad 1 \le j \le a$$
$$\hat{m}_{ij} + \hat{m}_{ji} = n_{ij} + n_{ji} \quad 1 \le i, j \le a$$

4.10.1   Relationships of Quasi Symmetry, Quasi Conditional Symmetry, and Quasi Odds Symmetry Models

The relationships of quasi symmetry (QS), quasi conditional symmetry (QCS), and quasi odds symmetry (QOS) models are depicted in Fig. 4.8. When $\gamma_i = \gamma$, QOS becomes QCS; when $\gamma = 1$, QCS becomes QS; and when $\gamma_j = 1$, QOS becomes QS (Fig. 4.18).

4.10.2   Relationships of Odds Symmetry I, Odds Symmetry II, and Quasi Odds Symmetry Model

As noted, QOS can be represented by either OS1 or OS2 when using GLM to generate the model. When we remove the two terms λ_i^R and λ_j^C from QOS, it reduces to either OS1 or OS2, depending on the specification of the QOS. The following show these associations (Fig. 4.19).

Fig. 4.18 Relationships of odds symmetry I, odds symmetry II, and quasi odds symmetry models

$$H_{QOS} : \pi_{ij}\pi_{jk}\pi_{ki} = \gamma_j\pi_{ji}\pi_{kj}\pi_{ik} \qquad H_{QCS} : \pi_{ij}\pi_{jk}\pi_{ki} = \gamma\pi_{ji}\pi_{kj}\pi_{ik}$$

QOS $\xrightarrow{\quad \gamma_i = \gamma \quad}$ QCS

$\gamma_j = 1$ $\qquad$ $\gamma = 1$

QS

$$H_{QS} : \pi_{ij}\pi_{jk}\pi_{ki} = \pi_{ji}\pi_{kj}\pi_{ik}$$

Fig. 4.19 Relationships of quasi symmetry, quasi conditional symmetry, and quasi odds symmetry models

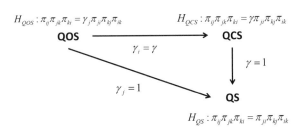

$$\ln(\hat{m}_{ij}) = \mu + \lambda_{ij}^S + \lambda_{ij}^{OS1}$$

OS1

$$\ln(\hat{m}_{ij}) = \mu + \lambda_i^R + \lambda_j^C + \lambda_{ij}^S + \lambda_{ij}^{OS1}$$

QOS $\qquad \lambda_i^R = \lambda_j^C = 0$

$$\ln(\hat{m}_{ij}) = \mu + \lambda_i^R + \lambda_j^C + \lambda_{ij}^S + \lambda_{ij}^{OS2}$$

OS2

$$\ln(\hat{m}_{ij}) = \mu + \lambda_{ij}^S + \lambda_{ij}^{OS2}$$

Exercises

4.1   Write a R function to generate the S factor for an $a \times a$ table using the following formula.

$$S_{ij} = \begin{cases} (k+1) - (i+1)(\frac{1}{2}i+1) + (a+3)(i+1) - 3 - 2a & i < j \\ (k+1) - (j+1)(\frac{1}{2}j+1) + (a+3)(j+1) - 3 - 2a & i \ge j \end{cases}$$

4.2   Write a R function to generate factor variables for OS1 and OS2 models using the following specification.

$$OS1_{ij} = \begin{cases} i & i < j \\ (a-1) + j & i > j \\ 2a - 1 & i = j \end{cases}$$

$$OS2_{ij} = \begin{cases} (2a - j) & i < j \\ (a+1) - i & i > j \\ 2a - 1 & i = j \end{cases}$$

4.3   Write a R function to generate the CS, D, and F factors for an $a \times a$ table using the following formula.

$$CS_{ij} = \begin{cases} 2 & i > j \\ 3 & i < j \\ 1 & i = j \end{cases}, \quad D_{ij} = \begin{cases} |i - j| & i < j \\ |i - j| + a - 1 & i > j, \\ 2a - 1 & i = j \end{cases}$$

$$F_{ij} = \begin{cases} |i - j| + 1 & i < j \\ 1 & \text{otherwise} \end{cases}$$

4.4   Fit complete symmetry, conditional symmetry, and quasi symmetry models for the following table. The data were obtained from a sample of 55,981 residences sampled by the U S Bureau of the Census on four regions. The four regions are Northeast, Midwest, South, and West of the USA (Table 4.15).

4.5   The following is the unaided distant vision of 4746 students of the Tokyo University about their left and right eye grade (Tomizawa 1985). Generate the asymmetry models discussed in this chapter and examine the fit of these asymmetry models (Table 4.16).

Table 4.15 Residence in 1980 and 1985

Residence in 1980	Residence in 1985			
	Northeast	Midwest	South	West
Northeast	11,607	100	366	124
Midwest	87	13,677	515	302
South	172	225	17,819	270
West	63	176	286	10,192

Table 4.16 Right and left eye grade

Right eye grade	Left eye grade				Total
	Lowest (1)	Second (2)	Third (3)	Highest (4)	
Lowest (1)	1429	249	25	20	1723
Second (2)	185	660	124	64	1033
Third (3)	23	114	221	149	507
Highest (4)	22	40	130	1291	1483
Total	1659	1063	500	1524	4746

4.6 The following is the unaided distant vision of 7477 British women from 1943 to 1946 reported in Staurt (1955). Generate the asymmetry models discussed in this chapter (Table 4.17).

4.7 Write a function to extract fit statistics for doubly classified model using package gnm, function gnm instead of function lm from the base. Include AIC BIC G^2, and the p-value.

4.8 Fit the following table with quasi conditional symmetry model, quasi symmetry model, and conditional symmetry model. Comment on the results (Table 4.18).

Table 4.17 Right and left eye grade

Right eye grade	Left eye grade				Total
	Best (1)	Second (2)	Third (3)	Worst (4)	
Best (1)	1520	266	124	66	1976
Second (2)	234	1512	432	78	2256
Third (3)	117	362	1772	205	2456
Worst (4)	36	82	179	492	789
Total	1907	2222	2507	841	7477

Table 4.18 Score by psychologists 1 and 2

Score by psychiatrist 2	Score by psychiatrist 1							
	1	2	3	4	5	6	7	8
1	372	30	10	11	20	20	14	20
2	1	650	10	5	4	3	1	1
3	1	5	840	22	3	5	3	5
4	1	4	4	1075	60	11	2	8
5	0	6	1	8	398	60	11	10
6	1	5	4	5	10	339	45	28
7	0	3	2	3	2	7	78	35
8	3	20	16	15	10	15	15	300

Table 4.19 Sweden elections in 1968 and 1970

1968	1970				Total
	Social democrat	Center	People's	Conservative	
Social democrat	850	35	25	6	916
Center	9	286	21	6	322
People's	3	26	185	5	219
Conservative	3	26	27	138	194
Total	865	373	258	155	1651

Table 4.20 Election in year 2013 and year 2017

Year 2013	Year 2017			
	People	Worker	Democratic	Labor
People	1520	267	133	57
Worker	233	1512	423	88
Democratic	109	370	1772	205
Labor	45	73	180	492

4.9   The results of a panel of voters in the election in Sweden in years 1968 and 1970 were tabulated below. The parties are arranged from left to right. Fit Table 4.19 with quasi symmetry model and comment on the results.

4.10   Fit quasi conditional symmetry model for Table 4.20.

References

Agresti, A. (1983). A simple diagonals-parameter symmetry and quasi-symmetry model. *Statistics and Probability Letters, 1*(6), 313–316.

Bhapkar, V. P. (1979). On tests of marginal symmetry and quasi-symmetry in two and three-dimensional contingency tables. *Biometrics, 35*(2), 417–426.

Bishop, Y. M., Fienberg, S. E., & Holland, P. W. (1975). *Discrete multivariate analysis: Theory and applications.* Springer.

Caussinus, H. (1965). Contribution à l'analyse statistique des tableaux de corrélation. *Annales de la faculté des sciences de Toulouse Sér. 4, 29*, 77–183.

Clogg, C. C., Eliason, S. R. & Grego, J. M. (1990). Models for the analysis of change in discrete variables, In A. von Eye (Ed.) *Statistical methods in longitudinal research* (Vol. 2, pp. 409–441). San Diego, CA: Academic Press.

Goodman, L. A. (1979a). Multiplicative models for square contingency tables with ordered categories. *Biometrika, 66*, 413–418.

Goodman, L. A. (1979b). Simple models for the analysis of association in cross-classifications having ordered categories. *Journal of American Statistical Association, 74*, 537–552.

Goodman, L. A. (1985). The analysis of cross-classified data having ordered and/or unordered categories: Association models, correlation models, and asymmetry models for contingency tables with or without missing entries. *The Annals of Statistics, 13*, 10–69.

Lawal, H. B. (2000). Implementing point-symmetry models for square contingency tables having ordered categories in SAS. *Journal of the Italian Statistical Society, 9*, 1–22.

Lawal, H. B. (2001). Modeling symmetry models in square contingency tables with ordered categories. *Journal of Statistical Computation and Simulation, 71,* 59–83.

Lawal, H. B. (2003). *Categorical data analysis with SAS and SPSS applications.* Lawrence Erlbaum Associates, Publishers.

Lawal, H. B. (2004). Using a GLM to decompose the symmetry model in square contingency tables with ordered categories. *Journal of Applied Statistics, 31*(3), 279–303.

Lawal, H. B. & Sundheim, R. (2002). Generating factor variables for asymmetry, non-independence and skew-symmetry models in square contingency tables with a SAS macro. *Journal of Statistical Software, 7*(8), 1–23.

McCullagh, P. (1978). A class of parametric models for the analysis of square contingency tables with ordered categories. *Biometrika, 65,* 413–418.

Tomizawa, S. (1985). Decompositions for odds-symmetry models in a square contingency table with ordered categories. *Journal of the Japan Statistical Society, 15,* 151–159.

Tomizawa, S. (1987). Decomposition for 2-ratios-parameter symmetry model in square contingency tables with ordered categories. *Biometrical Journal, 1,* 45–55.

Tomizawa, S. (1991). A multiplicative model imposed more restrictions on the quasi-symmetry model for the analysis of square contingency tables. *Biometrical Journal, 5,* 573–577.

Tomizawa, S. & Tahata, K. (2007). The analysis of symmetry and asymmetry: Orthogonality of decomposition of symmetry into quasi-symmetry and marginal symmetry for multi-way tables. *Journal de la Société Française de Statistque, 148,* 3–35.

von Eye, A. & Spiel, C. (1996). Standard and nonstandard log-linear symmetry models for measuring change in categorical variables. *American Statistician, 50,* 300–305.

Yamagushi, K. (1990). Some models for the analysis of asymmetric association in square contingency tables with ordered categories. *Sociological Methodology, 20,* 181–212.

Chapter 5
Point Symmetry Models

Specifically for a doubly classified $a \times a$ table, a point symmetry has the following probabilities

$$\pi_{i_1 i_2}, 1 \leq i_1, i_2 \leq a$$

with the following condition satisfied.

$$\pi_{i_1 i_2} = \pi_{i_1^* i_2^*} \quad 1 \leq i_1, i_2 \leq a$$

where $i^* = a + 1 - i$ and $j^* = a + 1 - j$. This symmetry property in reference to the center is presented as a symbolic table below. The center of the table is indicated with the word "center." For this 4×4 table, it implies that $\pi_{1,1} = \pi_{4,4}$, $\pi_{1,2} = \pi_{4,3}$, $\pi_{1,3} = \pi_{4,2}$ and so on. The different colors of circles in Fig. 5.1 show the equivalent probabilities in symmetry position in reference to the center. For instance, the filled red circle of cell (1,1) and cell (4,4) is in symmetrical position to the center of the table and their probabilities are equally likely. Similar coloring and pattern of circles for the rest of the table show equivalent in probability.

The following subsections introduce the various point symmetry models discussed in Tomizawa (1985, 1986a, c) and Lawal (2000). Similarly, the nonstandard log-linear model approach is used for modeling these point symmetry models.

5.1 Complete Point Symmetry Model

The complete point symmetry model is the simplest form of point symmetry model. It is the baseline model for the series of point symmetry models that will be discussed in this chapter. The hypothesis of the complete point symmetry model is stated below.

© Springer Nature Singapore Pte Ltd. 2017
T.K. Tan, *Doubly Classified Model with R*,
https://doi.org/10.1007/978-981-10-6995-6_5

Fig. 5.1 Point symmetry
model

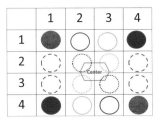

Fig. 5.2 Complete point
symmetry model

	1	2	3	4
1	π_{11}	π_{12}	π_{13}	π_{14}
2	π_{21}	π_{22}	π_{23}	π_{24}
3	π_{31}	π_{32}	π_{33}	π_{34}
4	π_{41}	π_{42}	π_{43}	π_{44}

$$H_p : \pi_{ij} = \pi_{i^*j^*}   1 \leq i \leq a;   1 \leq j \leq a$$

The symbolic table below describes the characteristics of complete point symmetry model. Those cells that with the same color and line pattern in circle are with
the same probability. It is observed that the cells with equivalent probabilities are in
reference to the center of the table. For instance, $\pi_{11} = \pi_{44}, \pi_{12} = \pi_{43}$ and so on.
(Fig. 5.2)

Point symmetry model (P) has the following log-linear model formulation.

$$\ln(\pi_{ij}) = \mu + \lambda_i^R + \lambda_j^C + \lambda_{ij}^{RC}$$

where

$$\sum_{i=1}^{a} \lambda_i^R = \sum_{j=1}^{a} \lambda_j^C = \sum_{i=1}^{a} \lambda_{ij}^{RC} = \sum_{j=1}^{a} \lambda_{ij}^{RC} = 0$$

$$\lambda_i^R = \lambda_{i^*}^R, \lambda_j^C = \lambda_{j^*}^C, \text{ and } \lambda_{ij}^{RC} = \lambda_{i^*j^*}^{RC}$$

The maximum likelihood of m_{ij} is simply the average of two cells in point
symmetry.

$$m_{ij} = \frac{n_{ij} + n_{i^*j^*}}{2}$$

Point symmetry model in nonstandard log-linear form is as follows,

$$l_{ij} = \mu + \lambda_{ij}^P$$

where P is a factor variable specified below. For a 4×4 table, the P factor variable is specified as follows:

$$P = \begin{pmatrix} 1 & 3 & 4 & 5 \\ 8 & 2 & 6 & 7 \\ 7 & 6 & 2 & 8 \\ 5 & 4 & 3 & 1 \end{pmatrix}$$

For a 6×6 table, the P factor variable is specified as follows:

$$P = \begin{pmatrix} 1 & 4 & 5 & 6 & 7 & 8 \\ 18 & 2 & 9 & 10 & 11 & 12 \\ 17 & 16 & 3 & 13 & 14 & 15 \\ 15 & 14 & 13 & 3 & 16 & 17 \\ 12 & 11 & 10 & 9 & 2 & 18 \\ 8 & 7 & 6 & 5 & 4 & 1 \end{pmatrix}$$

From the P factor, we can observe that the symmetry of the factor specifications is in symmetry from the center of the matrix. For the 6×6 table, cell (1,1) and cell (6,6) is symmetry to the center, specified as the number 1. Similarly, the cell (1,6) and cell (6,1) also in reference to the center of the matrix, specified as number 8. The rest of the cells follow this rule of specifying the factors.

For a 4×4 table, the model could also be re-specified as a generalized linear model with Poisson family since the dependent is a count of cell frequencies. Since there are eight factors, there are eight dummy variables specified. While generating the model, one dummy will be dropped, which generally is the first dummy, i.e., $\lambda_1^P = 0$.

$$l_{ij} = \mu + \lambda_1^S Z_{1ij} + \lambda_2^S Z_{2ij} + \lambda_3^S Z_{3ij} + \lambda_4^S Z_{4ij} + \lambda_5^S Z_{5ij} + \lambda_6^S Z_{6ij} + \lambda_7^S Z_{7ij} + \lambda_8^S Z_{8ij}$$

$$Z_{1ij} = \begin{cases} 1 & \text{cell } (1,1), \text{ cell } (4,4) \\ 0 & \text{otherwise} \end{cases}$$

$$Z_{2ij} = \begin{cases} 1 & \text{cell } (2,2), \text{ cell } (3,3) \\ 0 & \text{otherwise} \end{cases}$$

$$\cdots$$

$$Z_{7ij} = \begin{cases} 1 & \text{cell } (2,4), \text{ cell } (3,1) \\ 0 & \text{otherwise} \end{cases}$$

$$\lambda_1^P = 0$$

The nonstandard log-linear model is thus:

$$
\begin{bmatrix} l_{11} \\ l_{12} \\ l_{13} \\ l_{14} \\ l_{21} \\ l_{22} \\ l_{23} \\ l_{24} \\ l_{31} \\ l_{32} \\ l_{33} \\ l_{34} \\ l_{41} \\ l_{42} \\ l_{43} \\ l_{44} \end{bmatrix}
=
\begin{bmatrix}
1 & 1 & 0 & 0 & 0 & 0 & 0 & 0 & 0 \\
1 & 0 & 0 & 1 & 0 & 0 & 0 & 0 & 0 \\
1 & 0 & 0 & 0 & 1 & 0 & 0 & 0 & 0 \\
1 & 0 & 0 & 0 & 0 & 1 & 0 & 0 & 0 \\
1 & 0 & 0 & 0 & 0 & 0 & 0 & 0 & 1 \\
1 & 0 & 1 & 0 & 0 & 0 & 0 & 0 & 0 \\
1 & 0 & 0 & 0 & 0 & 0 & 1 & 0 & 0 \\
1 & 0 & 0 & 0 & 0 & 0 & 0 & 1 & 0 \\
1 & 0 & 0 & 0 & 0 & 0 & 0 & 1 & 0 \\
1 & 0 & 0 & 0 & 0 & 0 & 1 & 0 & 0 \\
1 & 0 & 1 & 0 & 0 & 0 & 0 & 0 & 0 \\
1 & 0 & 0 & 0 & 0 & 0 & 0 & 0 & 1 \\
1 & 0 & 0 & 0 & 0 & 1 & 0 & 0 & 0 \\
1 & 0 & 0 & 0 & 1 & 0 & 0 & 0 & 0 \\
1 & 0 & 0 & 1 & 0 & 0 & 0 & 0 & 0 \\
1 & 1 & 0 & 0 & 0 & 0 & 0 & 0 & 0
\end{bmatrix}
\begin{bmatrix} \mu \\ \lambda_1 \\ \lambda_2 \\ \lambda_3 \\ \lambda_4 \\ \lambda_5 \\ \lambda_6 \\ \lambda_7 \\ \lambda_8 \end{bmatrix}
$$

The above formulation is the familiar regression model with the design matrix consists of 0s and 1s, which is equivalent of the P factor variable specified above. For instance, the second column of 1s appears on cell (1,1) and cell (4,4) which is corresponding position of the P factor variable.

The following Table 5.1 shows the cross-tabulation of 626 students expressed their level of self-efficacy in a four-point scale measurement over two periods of time from a study.

The R syntax to generate the complete point symmetry model for Table 5.1 is stated below.

```
CPSData <- c(70, 15, 23, 81,
             28, 34, 10, 51,
             47, 11, 35, 30,
             76, 27, 15, 73)
P <- c(1, 3, 4, 5,
```

Table 5.1 Self efficacy of students—year 2016 × year 2017

Self-efficacy 2016	Self-efficacy 2017			
	1	2	3	4
1	70	15	23	81
2	28	34	10	51
3	47	11	35	30
4	76	27	15	73

```
        8, 2, 6, 7,
        7, 6, 2, 8,
        5, 4, 3, 1)
P <- as.factor(P)
CompletePoint <- glm(CPSData ~ P, family=poisson)
summary(CompletePoint)
model.summary(CompletePoint)
```

Alternatively, specified the *P* factor as separate vectors generate the same result using glm function specified as follows:

```
S11_44 <- c(1,0,0,0,0,0,0,0,0,0,0,0,0,0,0,1)
S12_43 <- c(0,1,0,0,0,0,0,0,0,0,0,0,0,0,1,0)
S13_42 <- c(0,0,1,0,0,0,0,0,0,0,0,0,0,1,0,0)
S14_41 <- c(0,0,0,1,0,0,0,0,0,0,0,0,1,0,0,0)
S21_34 <- c(0,0,0,0,1,0,0,0,0,0,0,1,0,0,0,0)
S22_33 <- c(0,0,0,0,0,1,0,0,0,0,1,0,0,0,0,0)
S23_32 <- c(0,0,0,0,0,0,1,0,0,1,0,0,0,0,0,0)
S24_31 <- c(0,0,0,0,0,0,0,1,1,0,0,0,0,0,0,0)
CompletePoint1 <- glm(CPSData ~ S11_44 + S22_33 + S12_43 + S13_42 +
                                S14_41 + S23_32 + S24_31, family=poisson)
```

The estimated G^2 is 0.84 with p-value 0.9991 shows that the complete point symmetry model fits well.

```
[1] "Model Summary:"
 Df    G2       p    AIC     BIC
  8 0.84 0.9991 -15.16 -50.68
```

The estimated values show the symmetry in reference to the center of the table. The fitted values of both cell (1,1) and cell (4,4) are 71.5 which is the average of the raw frequencies of cell (1,1) and cell (4,4) $[(70+73)/2]$. Other fitted values could also be observed in the same manner. As the symmetry is at the center of the table, the local odds ratios also exhibit symmetry from the center of the local odds ratio table. For instance, cell (1,1) and cell (3,3) of the local odds ratio table both have the estimated odds ratio of 9.92.

$$\hat{m}_{ij} = \begin{bmatrix} 71.5 & 15 & 25 & 78.5 \\ 29 & 34.5 & 10.5 & 49 \\ 49 & 10.5 & 34.5 & 29 \\ 78.5 & 25 & 15 & 71.5 \end{bmatrix} \quad \hat{\theta}_{ij} = \begin{bmatrix} 9.92 & 0.56 & 1.44 \\ 0.20 & 10.16 & 0.20 \\ 1.44 & 0.56 & 9.92 \end{bmatrix}$$

5.2 Inclined Point Symmetry Model

The inclined point symmetry model is a decomposition of the conditional symmetry model (McCullagh 1978). It has the following hypothesis:

$$H_{IP} : \pi_{ij} = \begin{cases} r\pi_{i^*j^*} & 1 \leq i \leq j \leq a \\ \pi_{i^*i^*} & i = j \end{cases}$$

where $r > 0$, $i^* = a + 1 - i$, and $j^* = a + 1 - j$. Similar to conditional symmetry, the probability of off-diagonal upper cells and off-diagonal lower cells is described as a constant term. The difference between conditional and inclined point symmetry model is that the reference of symmetry for conditional symmetry model is the diagonal of a table, whereas the inclined point symmetry is the center of the table. The diagonal cells of the inclined point symmetry model follow the complete point symmetry model that these cells are in symmetry to the center of the table. The symbolic table below shows the point symmetry of the diagonal cells with circles, $\pi_{11} = \pi_{44}$ and $\pi_{22} = \pi_{33}$. The off-diagonal cells are enclosed by two red triangles with the property that the total of the upper cells is a constant r to the lower diagonal cells. This property also applies to individual cell as shown below, and they are in point symmetry to the center of the table (Fig. 5.3).

$$r = \frac{\pi_{12}}{\pi_{43}} = \frac{\pi_{13}}{\pi_{42}} = \frac{\pi_{14}}{\pi_{41}} = \frac{\pi_{21}}{\pi_{34}} = \frac{\pi_{23}}{\pi_{32}} = \frac{\pi_{24}}{\pi_{31}}$$

Inclined point symmetry model in nonstandard log-linear form is stated below. The degrees of freedom for $a \times a$ table is $(a^2 - 3)/2$ and $(a^2 - 2)/2$, respectively, for table of odd and even row and column.

$$l_{ij} = \mu + \lambda_{ij}^P + \phi\Psi_{ij}$$

where the factor variable P is defined previously in point symmetry model and Ψ_{ij} is a regression variable defined as follows:

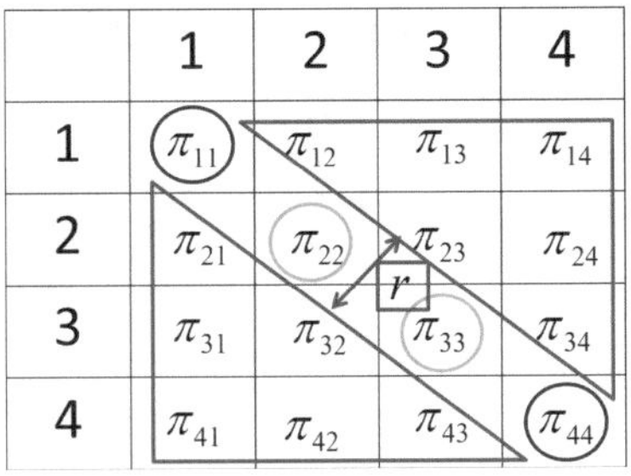

Fig. 5.3 Inclined point symmetry

$$\Psi_{ij} = \begin{cases} 1 & i<j \\ 2 & i>j \\ 3 & \text{otherwise} \end{cases}$$

For a 4×4 table, the Ψ factor is specified as follows:

$$\Psi = \begin{bmatrix} 3 & 1 & 1 & 1 \\ 2 & 3 & 1 & 1 \\ 2 & 2 & 3 & 1 \\ 2 & 2 & 2 & 3 \end{bmatrix}$$

The following Table 5.2 gives the tabulation of two doctors A and B about their opinions on the health status of 543 patients.

The R syntax to generate the inclined point symmetry model for the above table is stated below.

```
IPSData <- c(32, 17, 30, 83,
             29, 31, 12, 50,
             48, 15, 33, 30,
             63, 25, 16 29)
P  <- c(1, 3, 4, 5,
        8, 2, 6, 7,
        7, 6, 2, 8,
        5, 4, 3, 1)
P <- as.factor(P)
Psi <- c( 3, 1, 1, 1,
          2, 3, 1, 1,
          2, 2, 3, 1,
          2, 2, 2, 3)
InclinedPoint <- glm(IPSData ~ P + Psi, family=poisson)
summary(InclinedPoint)
model.summary(InclinedPoint)
```

The estimated values for the inclined point symmetry model are given below.

Table 5.2 Doctors opinion about health status of patients

Doctor A	Doctor B			
	1	2	3	4
1	32	17	30	83
2	29	31	12	50
3	48	15	33	30
4	63	25	16	29

128 5 Point Symmetry Models

```
> summary(InclinedPoint)

Call:
glm(formula = IPSData ~ P + Psi, family = poisson)

Deviance Residuals:
     Min        1Q    Median        3Q       Max
-0.66889  -0.24862   0.00335   0.25621   0.63873

Coefficients:
             Estimate Std. Error z value Pr(>|z|)
(Intercept)   3.79141    0.32071  11.822  < 2e-16 ***
P2            0.04801    0.17894   0.268  0.78847
P3           -0.80315    0.26309  -3.053  0.00227 **
P4           -0.29232    0.23895  -1.223  0.22119
P5            0.68395    0.21392   3.197  0.00139 **
P6           -1.00382    0.27559  -3.642  0.00027 ***
P7            0.28531    0.22163   1.287  0.19797
P8           -0.22212    0.23635  -0.940  0.34733
Psi          -0.12456    0.09801  -1.271  0.20377
---
Signif. codes:  0 '***' 0.001 '**' 0.01 '*' 0.05 '.' 0.1 ' ' 1
```

The fitted values and the odds ratios for the inclined point symmetry model are listed below. The estimated value of $r = 1.13$. This estimated r can be extracted from output of summary(InclinedPoint), Phi coefficient $[1/\exp(-0.12456)]$. The value of r can also be worked out from the estimated values as shown below.

$$\hat{m}_{ij} = \begin{bmatrix} 30.50 & 17.53 & 29.11 & 77.54 \\ 27.67 & 32 & 14.34 & 52.05 \\ 45.95 & 12.66 & 32 & 31.33 \\ 68.46 & 25.79 & 15.47 & 30.50 \end{bmatrix}$$

$$r = \frac{17.53}{15.47} = \frac{29.11}{25.79} = \frac{77.54}{68.46} = \frac{31.33}{27.67} = \frac{14.34}{12.66} = \frac{52.05}{45.95} = 1.13$$

In contrast to complete point symmetry model, the estimated odds ratios of inclined point symmetry model below show a symmetrical pattern of odds ratios that is not from the center of the odds ratio table.

$$\hat{\theta}_{ij} = \begin{bmatrix} 2.01 & 0.27 & 1.37 \\ 0.24 & 5.64 & 0.27 \\ 1.37 & 0.24 & 2.01 \end{bmatrix}$$

Relationship between Complete Point Symmetry and Inclined Point Symmetry

The relationship between complete point symmetry and inclined point symmetry is described by the diagram below. By specifying $r = 1$, inclined point symmetry reduces to complete point symmetry model. When $r = 1$, the constant relationship between the upper and lower diagonal cells disappears. So, it reduces to complete point symmetry of a constant probability of one (Fig. 5.4).

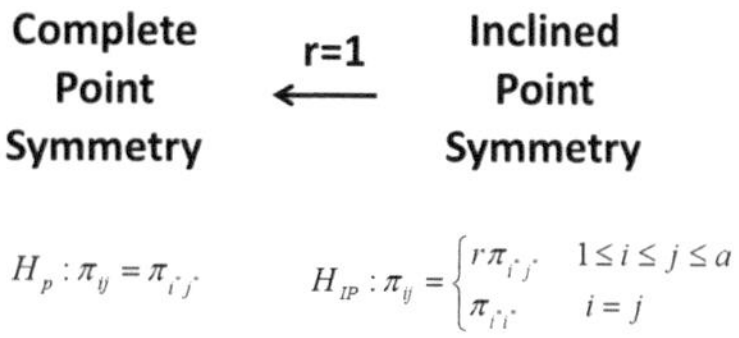

Fig. 5.4 Relationship
between complete point
symmetry and inclined point
symmetry model

$$\begin{array}{ccc} \textbf{Complete} & \xleftarrow{r=1} & \textbf{Inclined} \\ \textbf{Point} & & \textbf{Point} \\ \textbf{Symmetry} & & \textbf{Symmetry} \end{array}$$

$$H_p : \pi_{ij} = \pi_{i^*j^*} \qquad H_{IP} : \pi_{ij} = \begin{cases} r\pi_{i^*j^*} & 1 \le i \le j \le a \\ \pi_{i^*j^*} & i = j \end{cases}$$

5.3   Quasi Point Symmetry Model

While the complete point symmetry model is about symmetry in probabilities in reference to the center of the table, the quasi point symmetry model is about the odds ratios symmetry to the center. The hypothesis is specified as follows:

$$H_{Qp} : \frac{\pi_{il}/\pi_{jk}}{\pi_{il}/\pi_{jl}} = \frac{\pi_{j^*l^*}/\pi_{i^*l^*}}{\pi_{j^*l^*}/\pi_{i^*k^*}} \qquad 1 \le i < j \le a; \quad 1 \le k < l \le a$$

The quasi point symmetry model refers to a local odds ratios table that produces odds ratios that are symmetry to the center of the table. Put it in line with the above specification, the hypothesis H_{Qp} can be interpreted as the odds ratio produced from an arbitrary pair of rows where $i < j$ and an arbitrary pair of columns where $k < l$ is equivalent to that of the odds ratio produced from the point symmetric rows where $j^* < i^*$ for rows where $j > i$, respectively, and the point symmetric columns $l^{**} < k^{**}$ for columns $l > k$, respectively (Tomizawa 1985). Put in simply, there is a symmetry pattern of local odds ratios from the center for the entire table.

The symbolic table below restates the property of local odds ratio for quasi point symmetry model. The odds ratios with the same color are with the same estimated odds ratios. For instance, θ_{11} is the same as θ_{33}, θ_{12} is the same as θ_{32}, and so on. It is observed that the local odds ratios are in symmetry to the center of the local odds ratio table (Fig. 5.5).

The nonstandard log-linear model of quasi point symmetry model is similar to that of the quasi symmetry model with two additional factor specifications, R and C added to the baseline model. The difference between the two models is that the quasi symmetry model adds in baseline factor S, whereas quasi point symmetry adds in baseline factor P.

Fig. 5.5 Quasi point
symmetry model

	1	2	3
1	θ_{11}	θ_{12}	θ_{13}
2	θ_{21}	θ_{22}	θ_{23}
3	θ_{31}	θ_{32}	θ_{33}

Table 5.3 Life satisfaction of singaporean, Year 2016 × Year 2017

Life satisfaction year 2016	Life satisfaction year 2017			
	1	2	3	4
1	700	130	15	15
2	100	330	60	30
3	15	60	120	70
4	10	20	60	650

$$\ln\left(\pi_{ij}\right) = \mu + \lambda_i^R + \lambda_j^C + \lambda_{ij}^P$$

Table 5.3 tabulates the life satisfaction of respondents carried out in year the 2016 and 2017. A scale of 1 represents lowest satisfaction and 4 represents highest life satisfaction.

The R syntax to generate quasi point symmetry model for the above table is listed below.

```
QPData <- c(700, 130, 15, 15,
            100, 330, 60, 30,
            15, 60, 120, 70,
            10, 20, 60, 650)
r<-gl(4,4)
c<-gl(4,1,length=16)
P <- c(1, 3, 4, 5,
       8, 2, 6, 7,
       7, 6, 2, 8,
       5, 4, 3, 1)
P <- as.factor(P)
QuasiPoint <- glm(QPData ~ r + c + P, family=poisson)
summary(QuasiPoint)
model.summary(QuasiPoint)
```

The estimated values and odds ratios for quasi point symmetry model are shown below. Symmetry in odds ratio in reference to the center of the odds ratios table is observed. $\hat{\theta}_{11} = \hat{\theta}_{33} = 18.04$, $\hat{\theta}_{12} = \hat{\theta}_{23} = 1.54$, $\hat{\theta}_{13} = \hat{\theta}_{31} = 0.50$, and $\hat{\theta}_{21} = \hat{\theta}_{23} = 1.19$.

$$\hat{m}_{ij} = \begin{bmatrix} 700.52 & 129.27 & 15.17 & 15.04 \\ 99.37 & 330.84 & 59.94 & 29.85 \\ 15.15 & 60.06 & 119.16 & 70.63 \\ 9.96 & 19.83 & 60.73 & 649.48 \end{bmatrix} \quad \hat{\theta}_{ij} = \begin{bmatrix} 18.04 & 1.54 & 0.50 \\ 1.19 & 10.95 & 1.19 \\ 0.50 & 1.54 & 18.04 \end{bmatrix}$$

5.4 Quasi Inclined Point Symmetry Model

The quasi inclined point symmetry model is a decomposition of the inclined point symmetry model (Lawal 2000; Tomizawa 1986a). While the inclined point symmetry model is about a constant term on probabilities with symmetry cells, the quasi inclined point symmetry model is about a constant term on odds ratio of symmetry cells in reference to the center of the table. The quasi inclined point symmetry model has the following hypothesis (Tomizawa 1986a).

$$H_{QIP} : \pi_{ij} = \alpha_i \beta_j \rho_{ij} \quad 1 \le i, j \le a, \alpha_i > 0, \beta_j > 0, \rho_{ij} > 0,$$

and

$$\rho_{ij} = \begin{cases} \rho_{i^*i^*} & i = j \\ \phi \rho_{i^*j^*} & i < j \end{cases}$$

with

$$\prod_{h=1}^{a} \alpha_h = \prod_{h=1}^{a} \beta_h = 1$$

and

$$\prod_{h=1}^{a} \frac{\rho_{ih}}{\rho_{hj}} = \phi^{\alpha + 1 \pm (i+j)}, \quad \phi > 0$$

Alternatively, the hypothesis is restated below (Tomizawa 1986a).

$$H_{QIP} : \begin{cases} \dfrac{\pi_{ij}/\pi_{jj}}{\pi_{ik}/\pi_{jk}} = \phi \dfrac{\pi_{j^*k^*}/\pi_{i^*k^*}}{\pi_{j^*j^*}/\pi_{i^*j^*}} & 1 \le i < j < k \le a \\ \dfrac{\pi_{ii}/\pi_{ji}}{\pi_{ij}/\pi_{jj}} = \dfrac{\pi_{j^*j^*}/\pi_{i^*j^*}}{\pi_{j^*i^*}/\pi_{i^*i^*}} & 1 \le i < j \le a \end{cases}$$

The symbolic table below shows the main characteristics of quasi inclined point symmetry model. Those cells in circles are symmetry in odds ratios with reference to the center of the table. For instance, $\theta_{11} = \theta_{44}$, $\theta_{13} = \theta_{42}$, and $\theta_{14} = \theta_{41}$. Those cells in square are one position away from the diagonal with the characteristics of a ratio ϕ of odds ratios to the opposite diagonal in reference to the center of the table. For instance, $\theta_{12}\phi = \theta_{43}$, $\theta_{21}\phi = \theta_{34}$, and $\theta_{32}\phi = \theta_{23}$ (Fig. 5.6).

The nonstandard log-linear model formulation of quasi inclined point symmetry model, specified by Lawal (2000), is given below.

$$l_{ij} = \mu + \lambda_i^R + \lambda_j^C + \lambda_{ij}^P + \phi \Psi_{ij}$$

Fig. 5.6 Quasi inclined point symmetry model

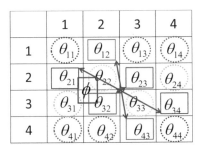

The factors R, C, and P and variable Ψ are defined previously.

A study was carried out asking respondents about their leadership and problem-solving skills in their job with 1 representing essentially not using it and 7 representing very frequently using it. The following Table 5.4 shows the tabulation of 395 workers who expressed their opinion on these two skills.

The R syntax to generate the quasi inclined point symmetry model for Table 5.4 is stated below.

```
QIPSData <- c(20, 10, 10, 13, 22, 21, 17,
              18, 11, 19, 12, 35, 32, 19,
              44,  9, 40, 22, 21, 13, 24,
              27,  9, 18, 35, 19, 14, 26,
              22, 10, 18, 22, 39, 12, 43,
              25, 25, 31, 11, 16, 17, 18,
              26, 21, 32, 18, 17, 21, 29)
r <- gl(7,7)
c <- gl(7,1,length=49)
Psi <- c( 3, 1, 1, 1, 1, 1, 1,
          2, 3, 1, 1, 1, 1, 1,
```

Table 5.4 Leadership and problem-solving indicators

Problem-solving skills	Leadership skills						
	1	2	3	4	5	6	7
1	20	10	10	13	22	21	17
2	18	11	19	12	35	32	19
3	44	9	40	22	21	13	24
4	27	9	18	35	19	14	26
5	22	10	18	22	39	12	43
6	25	25	31	11	16	17	18
7	26	21	32	18	17	21	29

```
              2, 2, 3, 1, 1, 1, 1,
              2, 2, 2, 3, 1, 1, 1,
              2, 2, 2, 2, 3, 1, 1,
              2, 2, 2, 2, 2, 3, 1,
              2, 2, 2, 2, 2, 2, 3)
P <- c( 1,  5,  6,  7,  8,  9, 10,
       25,  2, 11, 12, 13, 14, 15,
       24, 23,  3, 16, 17, 18, 19,
       22, 21, 20,  4, 20, 21, 22,
       19, 18, 17, 16,  3, 23, 24,
       15, 14, 13, 12, 11,  2, 25,
       10,  9,  8,  7,  6,  5,  1)
P <- as.factor(P)
QuasiInclinedPoint <- glm(QIPSData ~ r + c + P + Psi, family=poisson)
summary(QuasiInclinedPoint)
model.summary(QuasiInclinedPoint)
```

The estimated coefficients of quasi inclined point symmetry model are printed below.

```
            Estimate Std. Error z value Pr(>|z|)
(Intercept)  3.23077    0.49166   6.571 4.99e-11 ***
r2          -0.20693    0.31759  -0.652 0.514693
r3           0.71390    0.27486   2.597 0.009395 **
r4           0.20844    0.26962   0.773 0.439472
r5           0.69382    0.24835   2.794 0.005210 **
r6          -0.14976    0.27083  -0.553 0.580283
r7           0.45193    0.17086   2.645 0.008168 **
c2          -1.63880    0.26516  -6.180 6.40e-10 ***         P12   0.23458   0.32481   0.722 0.470156
c3          -0.42150    0.23264  -1.812 0.070014 .           P13   0.96646   0.29807   3.242 0.001185 **
c4          -0.73357    0.20486  -3.581 0.000342 ***         P14   1.88383   0.32375   5.819 5.93e-09 ***
c5          -0.40134    0.24584  -1.633 0.102566             P15   0.19987   0.22476   0.889 0.373862
c6          -1.33695    0.28053  -4.766 1.88e-06 ***         P16      NA        NA      NA      NA
c7          -0.10207    0.17258  -0.591 0.554238             P17  -0.44330   0.28816  -1.538 0.123948
P2           1.28074    0.42273   3.030 0.002448 **          P18   0.08724   0.30260   0.288 0.773113
P3           0.37534    0.34883   1.076 0.281928             P19  -0.63699   0.18231  -3.494 0.000476 ***
P4           1.07197    0.36589   2.930 0.003393 **          P20      NA        NA      NA      NA
P5           0.82644    0.38934   2.123 0.033783 *           P21   0.58458   0.34871   1.676 0.093654 .
P6          -0.35117    0.38779  -0.906 0.365157             P22      NA        NA      NA      NA
P7           0.11106    0.36835   0.302 0.763025             P23      NA        NA      NA      NA
P8           0.34574    0.36884   0.937 0.348564             P24      NA        NA      NA      NA
P9           1.18607    0.38830   3.054 0.002254 **          P25      NA        NA      NA      NA
P10         -0.25504    0.31826  -0.801 0.422916             Psi  -0.07408   0.15025  -0.493 0.621956
P11          0.33232    0.31979   1.039 0.298718             ---
                                                             Signif. codes:  0 '***' 0.001 '**' 0.01 '*' 0.05 '.' 0.1 ' ' 1
```

The fitted values and odds ratios are printed below. We observe that only those odds ratio that are one diagonal from the diagonal has a ratio ϕ of 0.92. For instance, $\theta_{12} \times 0.92 = \theta_{65}$. This ratio can be obtained from the estimated coefficient of Ψ, $\exp(-0.07408) = 0.92$.

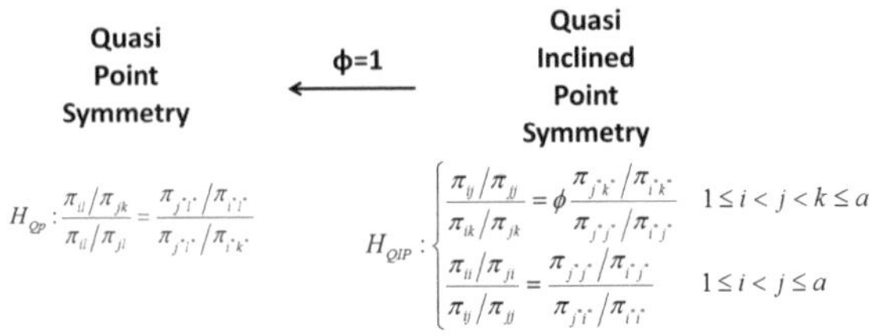

Fig. 5.7 Relationship between Quasi point symmetry and Quasi inclined point symmetry model

$$\phi = \frac{\hat{\theta}_{65}}{\hat{\theta}_{12}} = \frac{1.35}{1.46} = \frac{\hat{\theta}_{56}}{\hat{\theta}_{21}} = \frac{0.32}{0.30} = \frac{\hat{\theta}_{45}}{\hat{\theta}_{32}} = \frac{0.44}{0.41} = \frac{\hat{\theta}_{34}}{\hat{\theta}_{43}} = \frac{0.62}{0.57} = \frac{\hat{\theta}_{23}}{\hat{\theta}_{54}} = \frac{0.88}{0.82} = 0.92$$

$$\hat{m}_{ij} = \begin{bmatrix} 20.26 & 10.43 & 10.85 & 12.61 & 22.2 & 20.2 & 16.44 \\ 17.73 & 11.51 & 17.47 & 11.60 & 33.61 & 22.00 & 21.06 \\ 44.54 & 8.65 & 39.50 & 23.04 & 20.61 & 13.75 & 22/91 \\ 26.87 & 9.36 & 17.63 & 35 & 19.37 & 13.63 & 26.13 \\ 23.09 & 9.25 & 18.39 & 20.96 & 39.50 & 12.35 & 42.45 \\ 22.93 & 23.99 & 32.39 & 11.40 & 17.53 & 16.49 & 18.26 \\ 26.56 & 21.80 & 31.78 & 18.39 & 16.15 & 20.57 & 28.74 \end{bmatrix}$$

$$\hat{\theta}_{ij} = \begin{bmatrix} 1.26 & 1.46 & 0.57 & 1.64 & 1.08 & 0.78 \\ 0.30 & 3.01 & 0.88 & 0.31 & 0.68 & 2.61 \\ 1.79 & 0.41 & 3.40 & 0.62 & 1.06 & 1.15 \\ 1.15 & 1.06 & 0.57 & 3.40 & 0.44 & 1.79 \\ 2.61 & 0.68 & 0.31 & 0.82 & 3.01 & {}^{\backprime}0.32 \\ 0.78 & 1.08 & 1.64 & 0.57 & 1.35 & 1.26 \end{bmatrix}$$

Relationship between Quasi Point Symmetry and Quasi Inclined Point Symmetry

The relationship between quasi point symmetry and quasi inclined point symmetry is that the former is in total symmetry of local odds ratios in reference to the center of the local odds table, while the latter has restriction on a group of cells. This restriction refers to those cells that are one diagonal away from the diagonal is with a constant ϕ to the cells in reference to the center of the odds ratios table. As such, when $\phi = 1$, quasi inclined point symmetry model reduces to quasi point symmetry model (Fig. 5.7).

5.5 Proportional Point Symmetry Model

While the complete point symmetry model exhibits characteristic of complete symmetry in probability to the center of the table, the inclined point symmetry model exhibits the same property of complete point symmetry with the exception of those cells that are one position away from the diagonal that exhibits a constant r ratio in point symmetry. This chapter introduces another point symmetry model called proportional point symmetry model that also has the property of symmetry in probabilities with a ratio. However, the diagonal cells are not in point symmetry but remain the same values as the original data. The proportional point symmetry model has the following formulation.

$$H_{PP} : \pi_{ij} = \alpha \pi_{i^*j^*} \quad 1 \le i \le j \le a$$

The symbolic table for the proportional point symmetry model is shown in Fig. 5.6. The cells with the same colored circle are in symmetry. The diagonal cells are not circled, indicating the non-symmetry for these cells. The symbol α indicates the ratio of the symmetrical odds ratios (Fig. 5.8).

The proportional point symmetry model has degrees of freedom $(a+1)(a-2)/2$. The nonstandard log-linear formulation is as follows:

$$l_{ij} = \mu + \lambda_{ij}^{PP} + \alpha \Psi_{ij}$$

where Ψ_{ij} is a regression variable defined earlier and α is an unspecified regression parameter.

The general formula to generate the PP factor is stated below, but the formula does not provide the exact location of the factor for the off-diagonal cells.

$$PP_{ij} = \begin{cases} n_{ii} = i & i = j \\ n_{ij} == k & k = (a+1), (a+2), \ldots, \left[a + \binom{a}{2}\right] \end{cases}$$

For a 4×4 and a 6×6 contingency table, the above PP specification gives the following vector variable. The diagonal cells are with values running from 1 to i.

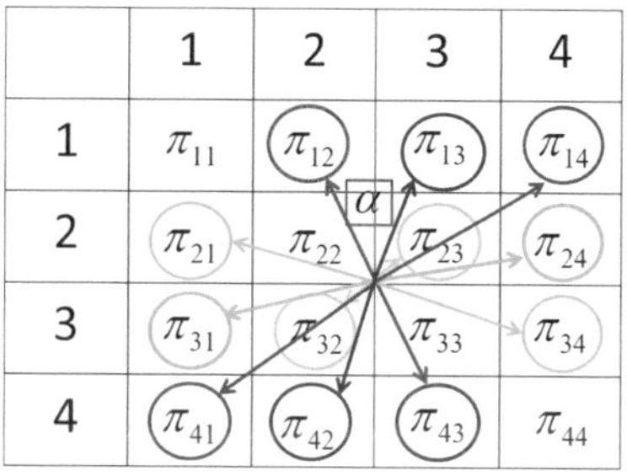

Fig. 5.8 Proportional point symmetry model

So, for a 4×4 and a 6×6 contingency table, it runs from 1 to 4 and 1 to 6, respectively. For the off-diagonal cells, it runs from 5 ($a + 1 = 4 + 1 = 5$) to 10 $\left(a + \begin{pmatrix} a \\ 2 \end{pmatrix} = 4 + 6 \right)$ for 4×4 table and from 7 ($a + 1 = 6 + 1$) to 21 $\left(a + \begin{pmatrix} a \\ 2 \end{pmatrix} = 6 + 15 = 21 \right)$ for a 6×6 table.

$$PP = \begin{pmatrix} 1 & 5 & 6 & 7 \\ 10 & 2 & 8 & 9 \\ 9 & 8 & 3 & 10 \\ 7 & 6 & 5 & 4 \end{pmatrix} \quad PP = \begin{pmatrix} 1 & 7 & 8 & 9 & 10 & 11 \\ 12 & 2 & 13 & 14 & 15 & 16 \\ 17 & 18 & 3 & 19 & 20 & 21 \\ 21 & 20 & 19 & 4 & 18 & 17 \\ 16 & 15 & 14 & 13 & 5 & 12 \\ 11 & 10 & 9 & 8 & 7 & 6 \end{pmatrix}$$

The following cross-tabulation summarizes the problem-solving skills and leadership skills of 395 workers.

The R syntax to generate the proportional point symmetry model for Table 5.5 is stated below.

```
PPSData <- c(81,  9,  5, 11,
             22,  8,  6, 17,
             33, 11, 43, 11,
             22, 11, 18, 87)
PP <- c( 1, 5, 6,  7,
        10, 2, 8,  9,
         9, 8, 3, 10,
         7, 6, 5,  4)
PP <- as.factor(PP)
Psi <- c( 3, 1, 1, 1,
          2, 3, 1, 1,
          2, 2, 3, 1,
          2, 2, 2, 3)
ProPoint <- glm(ParentEdu ~ PP + Psi, family=poisson)
summary(ProPoint)
model.summary(ProPoint)
```

Table 5.5 Problem-solving skills × leadership skills

Problem-solving skills	Leadership skills			
	1	2	3	4
1	81	9	5	11
2	22	8	6	17
3	33	11	43	11
4	22	11	18	87

The estimated values and the α value for the proportional point symmetry model are given below.

$$\hat{m}_{ij}$$

$$\hat{\alpha}$$

The $\hat{\alpha}$ could be found in the estimated coefficient of Phil from the summary (ProPoint) which is $\exp(0.68464) = 1.98$. It is noted the estimated values of the diagonal cells are the same of that of the original data.

5.5.1 Comparison of Proportional and Inclined Point Symmetry Model

The difference between proportional point symmetry and inclined point symmetry is the specification of the diagonal cells. The fit and the estimated coefficient of Phi for both the proportional and include point symmetry models for Table 5.5 are tabulated in Table 5.6. The fit statistics of proportional point symmetry are much better than the inclined point symmetry model. The difference in degrees of freedom for the two models is 2. The reason for the difference in df is that diagonal cells of inclined point symmetry model estimated two parameters while proportional point symmetry estimated four parameters. The diagonal cells for the former are symmetrical to the center while the latter is not. As such, the estimated values of the diagonal cells for proportional point symmetry are exactly the same as the raw data so there are no residuals for the diagonal cells. In contrast, the estimated values of the diagonal cells are averages of two cells. While the raw count of cell (2,2), 8, is far from the count of cell (3,3), 43. The residual is thus high for this cell. This explains why proportional point symmetry is better fit than inclined point symmetry. With regard to the proportion of upper and lower cells, the estimates for both models have exact estimates of 0.6846 up to four decimal places ($r = \alpha = 0.6846$). The selection between the models lies in the diagonal cells. If the diagonal cells have very different values and are not symmetrical, proportional point symmetry model is preferred to inclined point symmetry model.

Table 5.6 Comparison of proportional and inclined point symmetry models

Particular	Proportional point symmetry	Inclined point symmetry
G^2	0.07	26.67
df	5	7
p-value	0.9999	0.0004
AIC	−9.93	12.67
BIC	−29.83	−15.18
Estimated Phi coefficient	0.6846	0.6846

Fig. 5.9 Relationship between complete point symmetry, inclined point symmetry, and proportional point symmetry

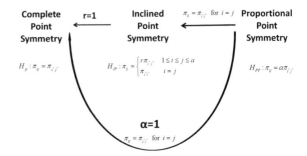

Relationship between Completed Point Symmetry Model, Inclined Point Symmetry Model, and Proportional Point Symmetry Model

The association of the three models, complete point symmetry, inclined point symmetry, and proportional point symmetry are described in the following diagram. As discussed earlier, when $r = 1$, inclined point symmetry reduces to complete point symmetry. When the specification of $\pi_{ij} = \pi_{i^*i^*}$ for $i = j$, the symmetry of diagonal cells, proportional point symmetry model reduces to inclined point symmetry. When an additional condition of $\alpha = 1$ is specified, proportional point symmetry reduces to complete point symmetry model (Fig. 5.9).

5.6 Local Point Symmetry Model

The local point symmetry model is yet another point symmetry model that exhibits point symmetry in probabilities. The difference between complete point symmetry and local point symmetry model is that the diagonal cells of the latter are not in point symmetry to the center of the table. The raw values of the diagonal cells remain the same as the fitted cell values. This model displays point symmetry with the exception for the observations when both row and column = i. The symbolic table below restates these properties with those cells in circle are symmetrical and the diagonal cells without circles enclosed are not in symmetry. The following states the hypothesis of local point symmetry model (Fig. 5.10).

$$H_L : \pi_{ij} = \pi_{i^*j^*} \quad i \neq j$$

Fig. 5.10 Local point symmetry model

	1	2	3	4
1	π_{11}	(π_{12})	π_{13}	(π_{14})
2	(π_{21})	π_{22}	(π_{23})	(π_{24})
3	(π_{31})	(π_{32})	π_{33}	(π_{34})
4	(π_{41})	π_{42}	(π_{43})	π_{44}

The nonstandard log-linear model formulation for local point symmetry model is as follows:

$$l_{ij} = \mu + \lambda_{ij}^{PP}$$

The degrees of freedom of local point symmetry model is $a(a-1)/2$. The factor PP for a 6×6 table is stated below.

$$PP = \begin{pmatrix} 1 & 7 & 8 & 9 & 10 & 11 \\ 12 & 2 & 13 & 14 & 15 & 16 \\ 17 & 18 & 3 & 19 & 20 & 21 \\ 21 & 20 & 19 & 4 & 18 & 17 \\ 16 & 15 & 14 & 13 & 5 & 12 \\ 11 & 10 & 9 & 8 & 7 & 6 \end{pmatrix}$$

Table 5.7 shows the right and left eye grades of 239 students when they go for a test on their eye sight.

The R syntax for fitting local point symmetry model for the above table is given below.

```
LPSData <- c(10, 16,  9, 17,
             23, 11,  7, 22,
             21,  8, 12, 25,
             18, 10, 17, 13)
PP <- c( 1,  5,  6,  7,
        10,  2,  8,  9,
         9,  8,  3, 10,
         7,  6,  5,  4)
PP <- as.factor(PP)
LocalPoint <- glm(LPSData ~ PP, family=poisson)
summary(LocalPoint)
model.summary(LocalPoint)
```

With the exception of the diagonal cells, the fitted values of the local point symmetry show the symmetrical pattern in reference to the center of the table. For instance, $\hat{m}_{12} = \hat{m}_{34} = 16.5$ and $\hat{m}_{13} = \hat{m}_{42} = 9.5$. The diagonal cells are with the same of the raw data values. For instance, $n_{11} = \hat{m}_{11} = 10$ and $n_{22} = \hat{m}_{22} = 11$.

Table 5.7 Right and left eye grade

Right eye grade	Left eye grade			
	1	2	3	4
1	10	16	9	17
2	23	11	7	22
3	21	8	12	25
4	18	10	17	13

$$\hat{m}_{ij} = \begin{bmatrix} 10 & 16.5 & 9.5 & 17.5 \\ 24 & 11 & 7.5 & 21.5 \\ 21.5 & 7.5 & 12 & 24 \\ 17.5 & 9.5 & 16.5 & 13 \end{bmatrix}$$

Reversal Point Symmetry Models

While the complete point symmetry, inclined point symmetry, quasi point symmetry, and proportional point symmetry are symmetrical to the center of table, another set of models referred to as reverse point symmetry models are based on a different diagonal symmetry specification. While the diagonal that mentioned in asymmetry models specifies the diagonal that runs from left top to right bottom, the reversal diagonal specifies the diagonal runs from right top to left bottom. Tomizawa (1985, 1986b) referred the set of point symmetry models that are based on both reversal diagonal symmetry and with point symmetry to the center of the table as the reverse point symmetry models. The following six reverse point symmetry models are discussed in the next six subsections

1. Reverse Local Point Symmetry Model,
2. Reverse Proportional Point Symmetry Model,
3. Reverse Inclined Point Symmetry Model,
4. Quasi Reverse Inclined Point Symmetry Model,
5. Reversed Conditional Point Symmetry Model,
6. Quasi Reversed Conditional Point Symmetry Model.

5.7   Reverse Local Point Symmetry Model

Tomizawa (1985) introduced the reverse local point symmetry model. As the name suggests, it is based on reverse diagonal and adds it on to the local point symmetry model. While the local point symmetry model shows symmetrical pattern in probability in reference to the center of a table with the exception of the diagonal cells (i, i), the exception of the reverse local point symmetry model refers to those cells that satisfy the condition $i + j = a + 1$. This condition refers to those cells at the diagonal that run from the upper right end to the lower bottom, the reverse diagonal. For a 4×4 table, the reverse diagonal cells include cell (1,4), cell (2,3), cell (3,2), and cell (4,1). These four cells satisfy the requirement that the sum of i and j is equal to $a + 1 = 5$. These are cells shown in the symbolic table below enclosed in reverse diagonal rectangular shape in orange color. The reverse diagonal here for the reverse local point symmetry model refers to the diagonal that runs from top left to bottom right. This is exactly the reversal of the local point symmetry that runs from top right to bottom left. The symbolic table shows those cells with circles remain unchanged in their point symmetry in respect to the center of the

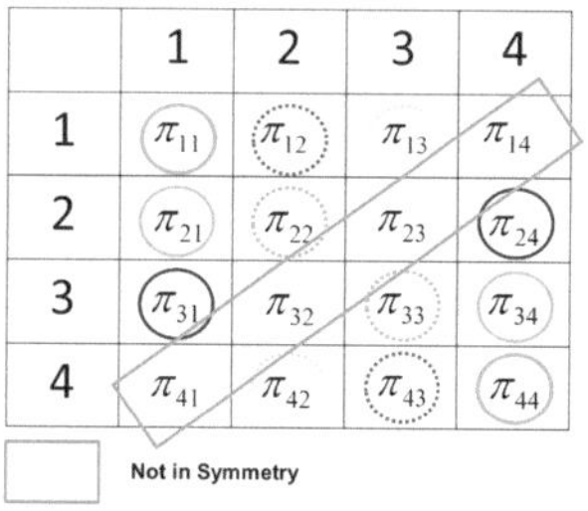

Fig. 5.11 Reverse local point symmetry model

table while the expectation of those cells in the rectangular shape remain unchanged, the reverse diagonal. The hypothesis for this model is thus specified as follows (Fig. 5.11).

$$H_R : \pi_{ij} = \pi_{i^*j^*} \quad i+j \neq a+1$$

The nonstandard log-linear model for reverse local point symmetry model has the following formulation with $a(a-1)/2$ degrees of freedom.

$$l_{ij} = \mu + \lambda_{ij}^{RL}$$

When a is even, RL factor is specified as follows:

$$RL_{ij} = \begin{cases} n_{ij} = \frac{a}{2} + k & k = 1, 2, \ldots \binom{a}{2} - 2 \\ n_{ij} = i & i = j \end{cases}$$

When a is odd, RL has the following specification.

$$RL_{ij} = \begin{cases} n_{ij} = \frac{a-1}{2} + k & k = 1, 2, \ldots \binom{a}{2} - 2 \\ n_{ij} = i & i = j \end{cases}$$

The RL factors for 4×4 and 6×6 table are shown below.

$$RL = \begin{pmatrix} 1 & 3 & 4 & 7 \\ 5 & 2 & 8 & 6 \\ 6 & 9 & 2 & 5 \\ 10 & 4 & 3 & 1 \end{pmatrix} \quad RL = \begin{pmatrix} 1 & 4 & 5 & 6 & 7 & 16 \\ 8 & 2 & 9 & 10 & 17 & 11 \\ 12 & 13 & 3 & 18 & 14 & 15 \\ 15 & 14 & 19 & 3 & 13 & 12 \\ 11 & 20 & 10 & 9 & 2 & 8 \\ 21 & 7 & 6 & 5 & 4 & 1 \end{pmatrix}$$

Table 5.8 shows the tabulation of 239 persons of their opinion which party to vote for in election year 2012 and 2016.

Table 5.8 Election
2012 × Election 2016

Election 2012	Election 2016			
	People	Worker	Liberal	Social
People	15	16	9	17
Worker	23	11	7	22
Liberal	21	8	11	25
Social	18	10	17	12

The *R* syntax below generates reverse local point symmetry model for the above
table.

```
RLPSData <- c(15, 16,  9, 17,
              23, 11,  7, 22,
              21,  8, 11, 25,
              18, 10, 17, 12)
RL <- c( 1, 3, 4, 7,
         5, 2, 8, 6,
         6, 9, 2, 5,
        10, 4, 3, 1)
RL <- as.factor(RL)
RevLocalPoint <- glm(RLPSData ~ RL, family=poisson)
summary(RevLocalPoint)
model.summary(RevLocalPoint)
```

The estimated values are shown below. It is observed that the reverse diagonal
cells are the same as the raw frequencies. For instance, $n_{14} = \hat{m}_{14} = 17$ and
$n_{23} = \hat{m}_{23} = 7$. The rest of the frequencies are in symmetrical to the center of the
table. For instance, $\hat{m}_{12} = \hat{m}_{43} = 16.5$ and $\hat{m}_{21} = \hat{m}_{34} = 24$.

$$\hat{m}_{ij} = \begin{bmatrix} 13.5 & 16.5 & 9.5 & 17 \\ 24 & 11 & 7 & 21.5 \\ 21.5 & 8 & 11 & 24 \\ 18 & 9.5 & 16.5 & 13.5 \end{bmatrix}$$

5.8   Reverse Proportional Point Symmetry Model

As mentioned in the previous section, diagonal cells refer to cells at diagonal that
run from top left to bottom right, whereas reverse diagonal cells refer to cells at
diagonal that run from top right to bottom left. This idea of reverse diagonal applies
not only to reverse local point symmetry model discussed in the last section, it also

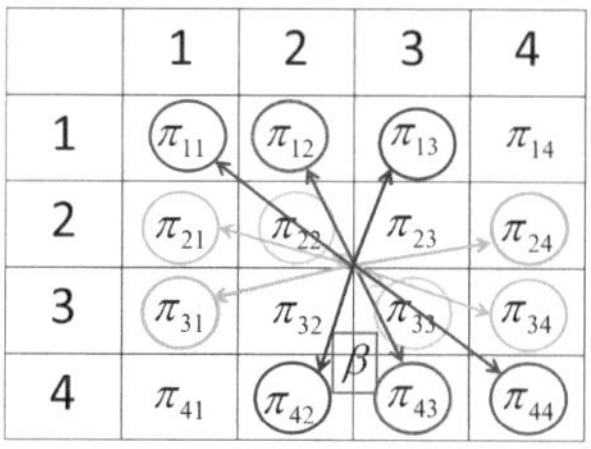

Fig. 5.12 Reverse
proportional point symmetry
model

applies to reverse proportional point symmetry model discussed in this section and the rest of the reverse point symmetry models. Under reverse local point symmetry model, the raw data and the fitted values of the reverse diagonal cells are with the same values. This property also applies to reverse proportional point symmetry model. The reverse diagonal cells in Fig. 5.12 are those cells not circled. The difference between reverse local point symmetry and reverse proportional point symmetry model is that the latter has an additional parameter β that shows the proportion of probability of the symmetrical upper and lower cells, whereas for reverse local point symmetry, this parameter β is equal to one. Thus, reverse proportional point symmetry model has the following hypothesis. As long as the cells are not reverse diagonal cells, i.e., when the condition $i+j<a+1$ satisfies, there exists a relationship between $\pi_{ij} = \beta\pi_{j^*i^*}$.

$$H_{RP} : \pi_{ij} = \beta\pi_{j^*i^*} \quad i+j<a+1$$

where the parameter β is unspecified.

The nonstandard log-linear model formulation for reverse proportional point symmetry model is as follows:

$$l_{ij} = \mu + \lambda_{ij}^{RP} + \beta\delta_{ij}$$

where RP is a factor variable which is equivalent to the factor variable RL defined in reverse point symmetry model. The RP factors for 4×4 and 6×6 table are shown below. The degrees of freedom for the model is $(a+1)(a-2)/2$.

$$RP = \begin{pmatrix} 1 & 3 & 4 & 7 \\ 5 & 2 & 8 & 6 \\ 6 & 9 & 2 & 5 \\ 10 & 4 & 3 & 1 \end{pmatrix} \quad RP = \begin{pmatrix} 1 & 4 & 5 & 6 & 7 & 16 \\ 8 & 2 & 9 & 10 & 17 & 11 \\ 12 & 13 & 3 & 18 & 14 & 15 \\ 15 & 14 & 19 & 3 & 13 & 12 \\ 11 & 20 & 10 & 9 & 2 & 8 \\ 21 & 7 & 6 & 5 & 4 & 1 \end{pmatrix}$$

The regression variable δ_{ij} is defined as follows:

$$\delta_{ij} = \begin{cases} 3 & i+j < a+1 \\ 1 & i+j = a+1 \\ 2 & \text{otherwise} \end{cases}$$

For a 4×4 and a 6×6 table, δ_{ij} have the following form:

$$\delta_{ij} = \begin{bmatrix} 3 & 3 & 3 & 1 \\ 3 & 3 & 1 & 2 \\ 3 & 1 & 2 & 2 \\ 1 & 2 & 2 & 2 \end{bmatrix}, \quad \delta_{ij} = \begin{bmatrix} 3 & 3 & 3 & 3 & 3 & 1 \\ 3 & 3 & 3 & 3 & 1 & 2 \\ 3 & 3 & 3 & 1 & 2 & 2 \\ 3 & 3 & 1 & 2 & 2 & 2 \\ 3 & 1 & 2 & 2 & 2 & 2 \\ 1 & 2 & 2 & 2 & 2 \end{bmatrix}$$

Table 5.9 shows the rating of 292 students by two raters A and B.

The R syntax to generate reverse proportional point symmetry model for the above table is shown below.

```
RProPData <- c(25, 34, 21, 17,
               51, 23,  7, 22,
               41, 18, 11, 25,
               58, 10, 17, 12)
RP <- c( 1, 3, 4, 7,
         5, 2, 8, 6,
         6, 9, 2, 5,
        10, 4, 3, 1)
RP <- as.factor(RP)
Delta <- c(3, 3, 3, 1,
           3, 3, 1, 2,
           3, 1, 2, 2,
           1, 2, 2, 2)
RevProPoint <- glm(RProPData ~ RP + Delta, family=poisson)
summary(RevProPoint)
model.summary(RevProPoint)
```

Table 5.9 Rater A × Rater B	Rater A	Rater B			
		1	2	3	4
	1	25	34	21	17
	2	51	23	7	22
	3	41	18	11	25
	4	58	10	17	12

The estimated values for reverse proportional point symmetry model and the estimated value of β are shown below.

$$\hat{m}_{ij} = \begin{bmatrix} 24.71 & 34.06 & 20.70 & 17 \\ 50.75 & 22.71 & 7 & 20.93 \\ 42.07 & 18 & 11.29 & 25.27 \\ 58 & 10.30 & 16.94 & 12.29 \end{bmatrix}$$

$$\hat{\beta} = \frac{34.06}{16.94} = \frac{20.70}{10.30} = \frac{22.71}{11.29} = \frac{50.75}{25.27} = \frac{42.07}{20.93} = 2.01$$

The estimated value of β can also be extracted from the estimated value of Psi $(\exp(0.69829) = 2.010312)$.

The estimated coefficients of reverse proportional point symmetry model are printed below.

```
Coefficients:
            Estimate Std. Error z value Pr(>|z|)
(Intercept)  1.11230    0.37000   3.006  0.00265 **
RP2         -0.08456    0.23757  -0.356  0.72189
RP3          0.32091    0.21595   1.486  0.13727
RP4         -0.17693    0.24349  -0.727  0.46744
RP5          0.71982    0.20046   3.591  0.00033 ***
RP6          0.53222    0.20712   2.570  0.01018 *
RP7          1.02263    0.35887   2.850  0.00438 **
RP8          0.13532    0.46133   0.293  0.76927
RP9          1.07979    0.35429   3.048  0.00231 **
RP10         2.24986    0.29531   7.619 2.56e-14 ***
Delta        0.69829    0.12425   5.620 1.91e-08 ***
---
Signif. codes:  0 '***' 0.001 '**' 0.01 '*' 0.05 '.' 0.1 ' ' 1
```

5.9   Reverse Inclined Point Symmetry Model

Tomizawa (1986b) introduced the reverse inclined point symmetry model. This third reverse point symmetry model follows the same rational of the first two reverse point symmetry models on reverse diagonal. The difference between reverse inclined point symmetry model and the previous two reverse point symmetry model is that the expected values of these cells are symmetrical to the center of the table instead of remaining the same value as the raw data. The symbolic diagram in

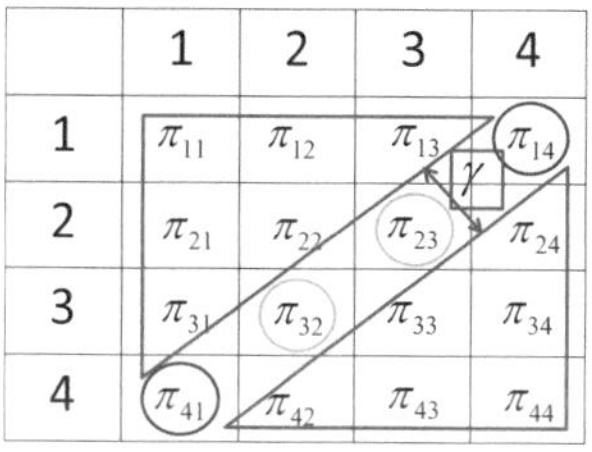

Fig. 5.13 Reverse inclined point symmetry model

Fig. 5.13 shows the pattern of symmetry by circle for the reverse diagonal cells. The rest of the cells are symmetrical to the center with a constant γ defined as follows for the reverse inclined point symmetry model.

$$H_{RIP} : \pi_{ij} = \begin{pmatrix} \gamma\pi_{i^*j^*} & i+j<a+1 \\ \pi_{ji} & i+j=a+1 \end{pmatrix}$$

where the parameter γ is unspecified

The nonstandard log-linear model formulation for reverse inclined point symmetry model is given by:

$$l_{ij} = \mu + \lambda_{ij}^{RI} + \gamma\delta_{ij}$$

where δ_{ij} is a regression parameter vector state in last section and restated below.

$$\delta_{ij} = \begin{cases} 3 & i+j<a+1 \\ 1 & i+j=a+1 \\ 2 & \text{otherwise} \end{cases}$$

For a 4×4 table, δ_{ij} has the following form:

$$\delta_{ij} = \begin{bmatrix} 3 & 3 & 3 & 1 \\ 3 & 3 & 1 & 2 \\ 3 & 1 & 2 & 2 \\ 1 & 2 & 2 & 2 \end{bmatrix}$$

The reverse inclined point symmetry model is based on $(a^2-2)/2$ and $(a^2-3)/2$ degrees of freedom for an even and odd table, respectively.

RI is a factor variable. For a 4×4 and a 6×6 table, the generated factor variables are as follows:

Table 5.10 Rater A × Rater B	Rater A	Rater B			
		1	2	3	4
	1	25	34	21	17
	2	51	23	7	22
	3	41	8	11	25
	4	16	10	17	12

$$RI = \begin{pmatrix} 1 & 3 & 4 & 7 \\ 5 & 2 & 8 & 6 \\ 6 & 8 & 2 & 5 \\ 7 & 4 & 3 & 1 \end{pmatrix} \quad RI = \begin{pmatrix} 1 & 4 & 5 & 6 & 7 & 16 \\ 8 & 2 & 9 & 10 & 17 & 11 \\ 12 & 13 & 3 & 18 & 14 & 15 \\ 15 & 14 & 18 & 3 & 13 & 12 \\ 11 & 17 & 10 & 9 & 2 & 8 \\ 16 & 7 & 6 & 5 & 4 & 1 \end{pmatrix}$$

Table 5.10 shows the ratings of 340 students by two raters A and B.

The R syntax to generate reverse inclined point symmetry model is shown below.

```
RIncPData <- c(25, 34, 21, 17,
               51, 23,  7, 22,
               41,  8, 11, 25,
               16, 10, 17, 12)
RI <- c( 1, 3, 4, 7,
         5, 2, 8, 6,
         6, 8, 2, 5,
         7, 4, 3, 1)
RI <- as.factor(RI)
Delta <- c(3, 3, 3, 1,
           3, 3, 1, 2,
           3, 1, 2, 2,
           1, 2, 2, 2)
RevIncPoint <- glm(RIncPData ~ RI + Delta, family=poisson)
summary(RevIncPoint)
model.summary(RevIncPoint)
```

The estimated values for reverse proportional point symmetry model and the estimated value of β are shown below.

$$\hat{m}_{ij} = \begin{bmatrix} 24.71 & 34.06 & 20.70 & 16.5 \\ 50.75 & 22.71 & 7.5 & 20.93 \\ 42.07 & 7.5 & 11.29 & 25.27 \\ 16.5 & 10.30 & 16.94 & 12.29 \end{bmatrix}$$

$$\hat{\gamma} = \frac{34.06}{16.94} = \frac{20.70}{10.30} = \frac{22.71}{11.29} = \frac{50.75}{25.27} = \frac{42.07}{20.93} = 2.01$$

Alternatively, the estimated value of γ can be extracted from the model estimated value of δ_{ij} (exp(0.69829) = 2.010312). The model output for reverse inclined point model is printed below.

```
Coefficients:
            Estimate Std. Error z value Pr(>|z|)
(Intercept)  1.11230    0.37000   3.006  0.00265 **
RI2         -0.08456    0.23757  -0.356  0.72189
RI3          0.32091    0.21595   1.486  0.13727
RI4         -0.17693    0.24349  -0.727  0.46744
RI5          0.71982    0.20046   3.591  0.00033 ***
RI6          0.53222    0.20712   2.570  0.01018 *
RI7          0.99277    0.31666   3.135  0.00172 **
RI8          0.20432    0.36964   0.553  0.58044
Delta        0.69829    0.12425   5.620 1.91e-08 ***
---
Signif. codes:  0 '***' 0.001 '**' 0.01 '*' 0.05 '.' 0.1 ' ' 1
```

5.10   Quasi Reverse Inclined Point Symmetry Model

Quasi reverse inclined point symmetry model is the mirror image of quasi inclined point symmetry model. While the latter is about symmetrical in odds ratios with the ratio ϕ for those cells one position away from the diagonal, quasi reverse inclined point symmetry model is with the ratio φ for those cells one position away from the reverse diagonal. These are cells with square shown in the symbolic table below. For those cells with circle, they are symmetrical in odds ratio in reference to the center of the odds ratios table. The quasi reverse inclined point symmetry model has the following formulation (Fig. 5.14)

$$H_{QRIP} : \begin{cases} \theta_{(i<j;k^*<j^*)} = \frac{1}{\varphi}\theta_{(j^*<i^*;j<k)} & 1\leq i<j<k\leq a \\ \theta_{(i<j;j^*<i^*)} = \theta_{(j^*<i^*;i<j)} & 1\leq i<j\leq a \end{cases}$$

where $\theta_{(i<j;k<l)} = \frac{\pi_{ik}\pi_{jk}}{\pi_{il}\pi_{jl}}$ for $1\leq i<j<k\leq a$

The quasi reverse inclined point symmetry model has the following nonstandard log-linear model formulation. The factors R, C, and RI and the regression variable Ψ are defined previously.

$$l_{ij} = \mu + \lambda_i^R + \lambda_j^C + \lambda_{ij}^{RI} + \varphi\delta_{ij}$$

The model is with degrees of freedom $(a^2 - 2a - 2)/2$ and $(a^2 - 2a - 1)/2$ for an even and odd table, respectively.

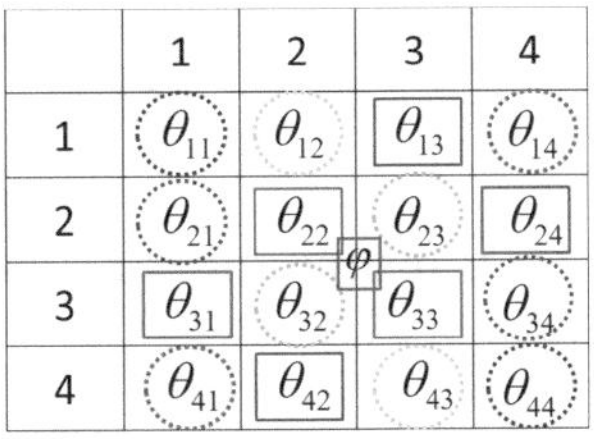

Fig. 5.14 Quasi reverse inclined point symmetry model

Table 5.11 Rater A × Rater B

Rater A	Rater B					
	1	2	3	4	5	6
1	25	35	32	25	22	16
2	33	23	15	20	22	20
3	29	14	11	22	13	51
4	56	14	30	12	15	27
5	26	22	16	12	18	31
6	16	21	21	28	33	25

The following table tabulates the assessment of rater A and B of 851 students, being 1 represents worse assessment rating and 6 represents best assessment rating (Table 5.11).

The *R* syntax to generate quasi reverse inclined point symmetry model for the above table is stated below.

```
QRIncPData <- c(25, 35, 32, 25, 22, 16,
                33, 23, 15, 20, 22, 20,
                29, 14, 11, 22, 13, 51,
                56, 14, 30, 12, 15, 27,
                26, 22, 16, 12, 18, 31,
                16, 21, 21, 28, 33, 25)
r <- gl(6,6)
c <- gl(6,1,length=36)
RI <- c( 1,  4,  5,  6,  7, 16,
         8,  2,  9, 10, 17, 11,
        12, 13,  3, 18, 14, 15,
        15, 14, 18,  3, 13, 12,
        11, 17, 10,  9,  2,  8,
        16,  7,  6,  5,  4,  1)
RI <- as.factor(RI)
Delta <- c(3, 3, 3, 3, 3, 1,
           3, 3, 3, 3, 1, 2,
           3, 3, 3, 1, 2, 2,
           3, 3, 1, 2, 2, 2,
           3, 1, 2, 2, 2, 2,
           1, 2, 2, 2, 2, 2)
QRevIncPoint <- glm(QRIncPData ~ r + c + RI + Delta, family=poisson)
summary(QRevIncPoint)
model.summary(QRevIncPoint)
```

The estimated values for quasi reverse inclined point symmetry model and the estimated value of φ are shown below.

$$\hat{\theta}_{ij} = \begin{bmatrix} 0.471 & 0.746 & 1.740 & 1.215 & 1.401 \\ 0.809 & 1.205 & 1.567 & 0.422 & 4.105 \\ 0.486 & 2.262 & 0.196 & 2.635 & 0.486 \\ 3.524 & 0.422 & 1.826 & 1.205 & 0.809 \\ 1.401 & 1.415 & 1.740 & 0.746 & 0.471 \end{bmatrix}$$

$$\hat{\varphi} = \frac{1.826}{1.567} = \frac{2.635}{2.262} = \frac{1.415}{1.215} = \frac{4.105}{3.524} = 1.1648$$

The estimated value of β can be extracted from the estimated value of Delta (exp (0.15258) = 1.1648). The estimated coefficients for quasi reverse inclined point model are printed below.

```
Coefficients: (4 not defined because of singularities)
            Estimate Std. Error z value Pr(>|z|)
(Intercept)  2.77620    0.60672   4.576 4.75e-06 ***
r2           1.50698    0.43282   3.482 0.000498 ***
r3           0.74073    0.20179   3.671 0.000242 ***
r4           0.83134    0.20258   4.104 4.06e-05 ***
r5           1.51195    0.48731   3.103 0.001918 **
r6           0.05825    0.18315   0.318 0.750459
c2          -1.35708    0.23052  -5.887 3.93e-09 ***
c3          -0.44629    0.38149  -1.170 0.242054
c4          -0.47044    0.39728  -1.184 0.236358
c5          -1.33738    0.25518  -5.241 1.60e-07 ***
c6           0.06397    0.18550   0.345 0.730212
RI2         -0.30151    0.34255  -0.880 0.378764
RI3         -1.04390    0.36630  -2.850 0.004374 **
RI4          1.68612    0.28487   5.919 3.24e-09 ***
RI5          0.67103    0.36966   1.815 0.069484 .
RI6          0.40647    0.37675   1.079 0.280643
RI7          1.22688    0.29964   4.095 4.23e-05 ***
RI8         -1.23424    0.39560  -3.120 0.001809 **
RI9         -1.60975    0.63701  -2.527 0.011503 *
RI10        -1.32029    0.62971  -2.097 0.036025 *
RI11        -1.56470    0.40548  -3.859 0.000114 ***
RI12        -0.64347    0.16507  -3.898 9.70e-05 ***
RI13         0.07747    0.26767   0.289 0.772266
RI14              NA         NA      NA       NA
RI15              NA         NA      NA       NA
RI16        -0.21730    0.34707  -0.626 0.531247
RI17              NA         NA      NA       NA
RI18              NA         NA      NA       NA
Delta        0.15258    0.17598   0.867 0.385916
---
Signif. codes:  0 '***' 0.001 '**' 0.01 '*' 0.05 '.' 0.1 ' ' 1
```

Fig. 5.15 Reverse conditional symmetry model

5.11 Reverse Conditional Symmetry

Reverse conditional symmetry also relied on reverse diagonal. However, the center is on the diagonal, not the center point of a table. The hypothesis of reverse conditional symmetry model is $\pi_{ij} = \eta\pi_{j^*i^*}$ for cells that satisfy the condition $i+j<a+1$. The symmetrical pattern is different from point symmetry model that specifies $\pi_{ij} = \pi_{i^*j^*}$ that is symmetrical to the center of the table. For instance, for a 4×4 table, cell (1,3) is symmetrical to cell (4,2) for point symmetry models , whereas reverse conditional symmetry the symmetrical pattern is cell (1,3) to cell (2,4). These symmetrical patterns are shown in the symbolic table below. Like the reverse models, the reverse diagonal cells are unaffected by the modeling, i.e., raw data are the same as the expected values. The specification of the reverse conditional symmetry model is as follows: (Fig. 5.15)

$$H_{RC} : \pi_{ij} = \eta\pi_{j^*i^*} \quad i+j<a+1$$

where the parameter η is unspecified.

The nonstandard log-linear model formulation is as follows:

$$l_{ij} = \mu + \lambda_{ij}^{RC} + \eta\delta_{ij}$$

where RC is the factor variable and δ_{ij} defined previously. For a 4×4, 5×5 and a 6×6 table, the RC factors are shown below. The reverse conditional symmetry model is based on $(a+1)(a-2)/2$ degrees of freedom.

$$RC = \begin{pmatrix} 1 & 3 & 4 & 7 \\ 5 & 2 & 8 & 4 \\ 6 & 9 & 2 & 3 \\ 10 & 6 & 5 & 1 \end{pmatrix}, \quad RC = \begin{pmatrix} 1 & 3 & 4 & 5 & 11 \\ 6 & 2 & 7 & 12 & 5 \\ 8 & 9 & 13 & 7 & 4 \\ 10 & 14 & 9 & 2 & 3 \\ 15 & 10 & 8 & 6 & 1 \end{pmatrix},$$

$$RC = \begin{pmatrix} 1 & 4 & 5 & 6 & 7 & 16 \\ 8 & 2 & 9 & 10 & 17 & 7 \\ 11 & 12 & 3 & 18 & 10 & 6 \\ 13 & 14 & 19 & 3 & 9 & 5 \\ 15 & 20 & 14 & 12 & 2 & 4 \\ 21 & 15 & 13 & 11 & 8 & 1 \end{pmatrix}$$

Table 5.12 below shows the ratings of 340 students by two raters A and B.

Rater A	Rater B			
	1	2	3	4
1	25	55	41	17
2	31	23	7	22
3	21	8	11	25
4	76	10	17	12

Table 5.12 Rater A × Rater B

```
RConPData <- c(25, 55, 41, 17,
               31, 23,  7, 22,
               21,  8, 11, 25,
               76, 10, 17, 12)


RC <- c( 1, 3, 4, 7,
         5, 2, 8, 4,
         6, 9, 2, 3,
        10, 6, 5, 1)
RC <- as.factor(RC)
Delta <- c(3, 3, 3, 1,
           3, 3, 1, 2,
           3, 1, 2, 2,
           1, 2, 2, 2)
RevConPoint <- glm(RConPData ~ RC + Delta, family=poisson)
summary(RevConPoint)
model.summary(RevConPoint)
```

The fitted values are shown below, and the estimated value of η is 2.02.

$$
\hat{m}_{ij} = \begin{bmatrix} 24.75 & 53.52 & 42.14 & 17 \\ 32.11 & 22.74 & 7 & 20.85 \\ 20.74 & 8 & 11.26 & 26.48 \\ 76 & 10.26 & 15.89 & 12.25 \end{bmatrix}
$$

$$
\eta = \frac{24.75}{12.25} = \frac{22.74}{11.26} = \frac{42.14}{20.85} = \frac{53.52}{26.48} = \frac{32.11}{15.89} = \frac{20.74}{10.26} = 2.02
$$

The estimated value of η could also be extracted from the estimated value of Delta ($\exp(0.7034) = 2.020611$).

```
Coefficients:
            Estimate Std. Error z value Pr(>|z|)
(Intercept)  1.09865    0.36987   2.970 0.002975 **
RC2         -0.08456    0.23757  -0.356 0.721894
RC3          0.77111    0.19881   3.879 0.000105 ***
RC4          0.53222    0.20712   2.570 0.010183 *
RC5          0.26028    0.21877   1.190 0.234142
RC6         -0.17693    0.24349  -0.727 0.467435
RC7          1.03116    0.35885   2.873 0.004060 **
RC8          0.14386    0.46131   0.312 0.755160
RC9          0.27739    0.44154   0.628 0.529848
RC10         2.52868    0.28829   8.771  < 2e-16 ***
Delta        0.70340    0.12414   5.666 1.46e-08 ***
---
Signif. codes:  0 `***' 0.001 `**' 0.01 `*' 0.05 `.' 0.1 ` ' 1
```

5.12 Quasi Reverse Conditional Symmetry Model

Lawal (2000) proposed the quasi reverse conditional symmetry model with the following hypothesis.

$$H_{QRC} : \theta_{(i<j; k^*<j^*)} = \frac{1}{\varphi} \theta_{(j<k; j^*<i^*)} \quad 1 \le i < j < k \le a$$

The difference between quasi reverse conditional symmetry model and quasi reverse inclined point symmetry model is that the reverse diagonal cells for the former are not symmetrical while the latter are symmetrical to the center of the table (Fig. 5.16).

The quasi-reverse conditional symmetry model has the following nonstandard log-linear model. It has $a(a-3)/2$ degrees of freedom. The various factors specified are the same as those mentioned earlier.

$$l_{ij} = \mu + \lambda_i^R + \lambda_j^C + \lambda_{ij}^{RC} + \varphi \delta_{ij}$$

832 students went through a mathematical test. The following table tabulates the ratings of rater A and B for these students, being 1 representing worse assessment rating and 6 representing best assessment rating (Table 5.13).

The R syntax for generating the quasi reverse conditional symmetry model for the above table is as follows:

```
QRConPData <- c(31, 51, 32, 24, 18, 20,
                31, 23, 11, 21,  2, 32,
```

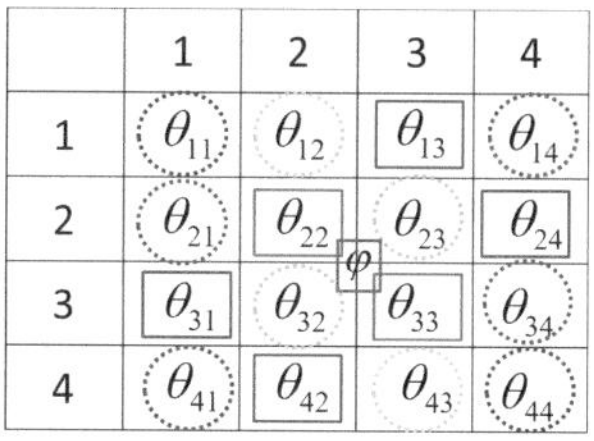

Fig. 5.16 Quasi reverse conditional symmetry model

Table 5.13 Rater A × Rater B

Rater A	Rater B					
	1	2	3	4	5	6
1	31	51	32	24	18	20
2	31	23	11	21	2	32
3	21	15	7	25	19	23
4	66	10	17	16	17	42
5	17	42	11	17	18	35
6	11	20	43	19	11	12

```
                    21, 15,  7, 25, 19, 23,
                    66, 10, 17, 16, 17, 42,
                    17, 42, 11, 17, 18, 35,
                    11, 20, 43, 19, 11, 12)
r <- gl(6,6)
c <- gl(6,1,length=36)
RC <- c( 1,  4,  5,  6,  7, 16,
         8,  2, 12, 13, 17,  7,
         9, 14,  3, 18, 13,  6,
        10, 15, 19,  3, 12,  5,
        11, 20, 15, 14,  2,  4,
        21, 11, 10,  9,  8,  1)
RC <- as.factor(RC)
Delta <- c(3, 3, 3, 3, 3, 1,
           3, 3, 3, 3, 1, 2,
           3, 3, 3, 1, 2, 2,
           3, 3, 1, 2, 2, 2,
           3, 1, 2, 2, 2, 2,
           1, 2, 2, 2, 2, 2)
QRevConPoint <- glm(QRConPData ~ r + c + RC + Delta, family=poisson)
summary(QRevConPoint)
model.summary(QRevConPoint)
```

The estimated values of the model showed the symmetrical patterns of local odds ratio for cells that are two positions away from the diagonal while the cells that are one position way from the diagonal is with a ratio of φ for the odds ratios.

Table 5.14 Happiness of Teachers 2015 and 2016

Happiness 2015	Happiness 2016					
		1	2	3	4	5	6
	1	75	77	36	16	5	1
	2	19	80	37	20	9	4
	3	10	40	71	30	23	1
	4	4	23	30	71	40	10
	5	3	9	2	31	80	20
	6	1	6	16	30	77	75

Table 5.15 Rating performance of prime minister

Primary survey	Second survey			
	1	2	3	4
1	40	5	11	10
2	15	8	8	12
3	35	9	45	21
4	15	6	22	42

Table 5.16 Leadership and problem-solving indicators

Problem-solving skills	Leadership skills				
	1	2	3	4	5
1	27	15	10	16	22
2	20	16	11	10	33
3	30	15	45	14	27
4	33	8	16	35	16
5	20	14	11	16	32

Table 5.17 Occupational status for Danish father–son Pairs

Father's status	Son's status				
	(1)	(2)	(3)	(4)	(5)
(1)	18	17	16	4	2
(2)	24	105	109	59	21
(3)	23	84	289	217	95
(4)	8	49	175	348	198
(5)	6	8	69	201	246

$$\hat{\theta}_{ij} = \begin{bmatrix} 0.49 & 0.79 & 2.25 & 0.14 & 14.24 \\ 0.76 & 0.95 & 2.32 & 8.00 & 0.07 \\ 0.25 & 3.43 & 0.25 & 1.21 & 2.26 \\ 15.96 & 0.15 & 1.78 & 0.95 & 0.79 \\ 0.74 & 8.29 & 0.25 & 0.76 & 0.49 \end{bmatrix}$$

$$\hat{\varphi} = \frac{0.07}{0.14} = \frac{1.21}{2.32} = \frac{1.78}{3.43} = \frac{8.29}{15.96} = 0.5193$$

The estimated value of φ could also be extracted from the estimated value of Delta ($\exp(-0.6552) = 0.5193$). The R output is not printed.

Exercises

5.1 The following table shows the cross-tabulation of responses of teachers expressing their happiness in a longitudinal study carried out in the year 2015 and 2016. Fit complete point symmetry model and comment (Table 5.14).

5.2 The following table below tabulates 304 respondents from two con-
secutive surveys before and after the prime minister implements a policy.
Fit a quasi inclined point symmetry model (Table 5.15).

5.3 A survey was carried out to ask respondents about their leadership and
problem-solving skills in their job with 1 presenting essentially not using
it 5 very frequently using it. The following table shows the
cross-tabulation of 395 workers who expressed their opinion on these
two skills. Fit a proportional point symmetry model and comment
(Table 5.16).

5.4 The following table is a 5 × 5 table of Danish social mobility data
(Bishop et al. 1975). Fit the various point symmetry models discussed in
this chapter and comment (Table 5.17).

References

Bishop, Y. M., Fienberg, S. E., & Holland, P. W. (1975). *Discrete multivariate analysis: Theory and applications*. Springer.

Lawal HB (2000) Implementing point-symmetry models for square contingency tables having ordered categories in SAS. {\it Journal of the Italian Statistical Society 9}, 1–22

McCullagh. (1978). A class of parametric models for the analysis of square contingency tables with ordered categories. {\it Biometrika 65}, 413–418.

Tomizawa, S. (1985). The decomposition for the inclined point-symmetry model in twoway contingency tables. {\it Biometrical Journal 27}, 895–905

Tomizawa, S. (1986a). A decomposition for the inclined point-symmetry model in a square contingency table. {\it Biometrical Journal 3}, 371–380

Tomizawa, S. (1986b). Four kinds of symmetry models and their decompositions in a square contingency table with ordered categories. {\it Biometrical Journal 28}, 387–393

Tomizawa, S. (1986c). The decompositions for point symmetry models in two-way contingency tables. {\it Biometrical Journal 8}, 895–905

Wall, K.D. & Lienert, G.A. (1976). A test for point-symmetry in J-dimensional contingency cubes. {\it Biometrische Zeitschrift.18}, 259–264

Chapter 6
Non-independence Models

The above classification of non-independence models into the three groups is based on the criteria listed in Table 6.1 (Upton 1985; Lawal 2003). The main characteristic of principal diagonal model is that the odds ratio is equal to one for the entire local odds ratio table with the exception of the diagonal cells. This is equivalently to state that the log odds ratio, Φ_{ij}, is equal to zero for the entire local log odds ratio table with the exception of the diagonal [When $\theta_{ij} = 1$, $\Phi_{ij} = 0$, given $\Phi_{ij} = \log(\theta_{ij})$]. Similar to principal diagonal, all the off-diagonal of odds ratio is equal to one; however, the exception is extended to include not only those cells on the diagonal but also cells that are one position away from the diagonal, i.e., $(|i - j| \leq 1)$. With regard to the characteristics of the odds ratio, the names of the model imply its characteristics. The principal diagonal refers to the main diagonal while the diagonal band extends to a band with cells that are one position away from the diagonal, and the full diagonal model has no restriction on the value of odds ratios need to be one.

6.1   Independence and Non-independence Model

Before proceed to discuss the various non-independence models, this section introduces the independence model, the null model for non-independence models. For an $a \times a$ contingency table with count m_{ij} in the ith and jth column of the table, the specification of the independence model is as follows:

$$m_{ij} = \alpha_i \beta_j \quad \text{for } i = 1, 2, \ldots, A; \quad j = 1, 2, \ldots, A$$

Similarly, for an $a \times a$ contingency table with r row variable denoted by R and c column variable denoted by C, the log-linear model is written as follows:

© Springer Nature Singapore Pte Ltd. 2017
T.K. Tan, *Doubly Classified Model with R*,
https://doi.org/10.1007/978-981-10-6995-6_6

Table 6.1 Criteria for the classification of non-independence models

Model	Criteria
Principal diagonal	$\Phi_{ij} = 0$ or $\theta_{ij} = 1$ unless $i = j$
Diagonal band	$\Phi_{ij} = 0$ or $\theta_{ij} = 1$ unless $|i - j| \leq 1$
Full diagonal	Need not have zero but preserve features of the original structural of the simpler models

$$\ell_{ij} = \ln(m_{ij}) = \mu + \lambda_i^R + \lambda_j^C + \lambda_{ij}^{RC}$$

where λ_{ij}^{RC} is referred as the interaction term. When $\lambda_{ij}^{RC} = 0$, it is referred to as the model of independence.

Table 6.2 tabulates the life satisfaction of parent and child.

The following R syntax fits the independent model for the above table.

```
IndepData <- c(4,   6,  11,  11,  20,   9,
               4,   6,  12,  11,  20,   8,
               9,  15,  25,  24,  46,  19,
              15,  22,  40,  39,  74,  30,
              17,  26,  45,  44,  83,  35,
              24,  35,  61,  60, 113,  48)
IndepData
r <- gl(6,6)
c <- gl(6,1,length=36)
r <- as.factor(r)
c <- as.factor(c)
Independent <- glm(IndepData ~ r + c, family=poisson)
summary(Independent)
model.summary(Independent)
```

The main characteristic of an independence model is the odds ratios that are exactly equal to one. The following symbolic table shows this property (Fig. 6.1).

Table 6.2 Life satisfaction of parent and child

Life satisfaction of parent	Life satisfaction of child					
	1	2	3	4	5	6
1	4	6	11	11	20	9
2	4	6	12	11	20	8
3	9	15	25	24	46	19
4	15	22	40	39	74	30
5	17	26	45	44	83	35
6	24	35	61	60	113	48

Fig. 6.1 Independence
model

θ	1	2	3	4
1	1	1	1	1
2	1	1	1	1
3	1	1	1	1
4	1	1	1	1

Non-independence Model

Independence model that fits social phenomena probably is rare and hard to come by. It is usually referred to as the baseline model for the various independence models which going to discuss in this chapter. Non-independence models are models that move away from independence. A non-independence model assumes the independence model as the baseline, and the interaction structure λ_{ij}^{RC} is modeled in some way. Often, λ_{ij}^{RC} is modeled as a function of local odds ratios θ_{ij} or log odds ratios, $\Theta_{ij} = \log(\theta_{ij}/\theta_{ji})$. In this context, the independence model is often referred to as the null or baseline model for the non-independence models. In the literature of mobility, the independence model is often referred to as perfect mobility (Goodman 1979b).

When the calculation of the odds ratio of composite adjacent 2×2 subtables to form $a\,(a-1) \times (a-1)$ table from $a \times a$ contingency table, the odds ratio θ_{ij} and the log odds ratios Φ_{ij} are defined as follows:

$$\hat{\theta}_{ij} = \left(\hat{m}_{ij}\hat{m}_{i+1,j+1}\right)/\left(\hat{m}_{i,j+1}\hat{m}_{i+1,j}\right)$$

$$\hat{\Phi}_{ij} = \ln\left(\hat{\theta}_{ij}\right), \quad \text{and}$$

$$\hat{\Theta}_{ij} = \ln\left(\hat{\theta}_{ij}\Big/\hat{\theta}_{ji}\right)$$

Generally, λ_{ij}^{RC} is modeled as a function of the local odds ratios θ_{ij} or Θ_{ij}. The log-linear model can thus be written in the following form:

$$\ell_{ij} = \ln\left(m_{ij}\right) = \mu + \lambda_i^R + \lambda_j^C + \Phi_{ij}$$

For the various models in this chapter, the log odds ratio $\hat{\Phi}_{ij} = \ln\left(\hat{\theta}_{ij}\right)$ has a diagonal pattern. For the independence model (O), $\hat{\Phi}_{ij} = 0$ for $(i,j) = 1, 2, \ldots, (I-1)$.

6.2 Principal Diagonal Models

Principal diagonal models have the following properties:

$$\Phi_{ij} = 0 \text{ or } \theta_{ij} = 1 \text{ unless } i = j$$

$$\text{where} \quad \begin{matrix} \Phi_{ij} = \ln(\theta_{ij}) \\ \theta_{ij} : \text{odds ratio} \end{matrix}$$

$$\Phi_{ij} = \begin{cases} 0 & i \neq j \\ \varphi_i & i = j \end{cases}$$

Two principal diagonal models are going to be discussed in this section, fixed distance and variable distance models (Haberman 1979; Lawal 2001, 2004).

6.2.1 Fixed Distance Model

While referring the parameter δ as the cells that are not on the main diagonal, the fixed distance model for an $a \times a$ table has the following multiplicative form (Goodman 1972, 1979b; Haberman 1974). The symbolic table below shows the main characteristic of fixed distance model that the entire odds ratios table is with odds ratios of one with the exception of the diagonal cells of a constant odds ratio θ (Fig. 6.2).

$$m_{ij} = \alpha_i \beta_j \delta^k \quad \text{for } k = |i - j|$$

The nonstandard log-linear model formulation for fixed distance model is stated below.

$$\ell_{ij} = \mu + \lambda_i^R + \lambda_j^C + \varphi F_{ij}$$

where F is a regression variable. For a 5×5 table, F is as follows:

Fig. 6.2 Fixed distance model

	1	2	3	4
1	θ	1	1	1
2	1	θ	1	1
3	1	1	θ	1
4	1	1	1	θ

$$F = \begin{pmatrix} 1 & 2 & 3 & 4 & 5 \\ 1 & 1 & 2 & 3 & 4 \\ 1 & 1 & 1 & 2 & 3 \\ 1 & 1 & 1 & 1 & 2 \\ 1 & 1 & 1 & 1 & 1 \end{pmatrix}$$

Table 6.3 tabulates the occupation status of grandfather and grandson.

The R syntax to generate fixed distance model for the above table is as follows:

```
GrSonGrFather <- c(22,  42,  20,  30,  16,
                   34, 180,  83, 130,  67,
                   17,  89, 120, 192, 100,
                   23, 128, 172, 785, 401,
                    9,  48,  64, 294, 435)
GrSonGrFather
r <- c(1, 1, 1, 1, 1,
       2, 2, 2, 2, 2,
       3, 3, 3, 3, 3,
       4, 4, 4, 4, 4,
       5, 5, 5, 5, 5)
r <- as.factor(r)
c <- c(1, 2, 3, 4, 5,
       1, 2, 3, 4, 5,
       1, 2, 3, 4, 5,
       1, 2, 3, 4, 5,
       1, 2, 3, 4, 5)
c <- as.factor(c)
F <- c(1, 2, 3, 4, 5,
       1, 1, 2, 3, 4,
       1, 1, 1, 2, 3,
       1, 1, 1, 1, 2,
       1, 1, 1, 1, 1)
FixedVariable <- glm(GrSonGrFather ~ r + c + F, family=poisson)
summary(FixedVariable)
model.summary(FixedVariable)
```

Table 6.3 Occupation status of grandfather and grandson

Grandfather's status	Grandson's status				
	(1)	(2)	(3)	(4)	(5)
(1)	22	42	20	30	16
(2)	34	180	83	130	67
(3)	17	89	120	192	100
(4)	23	128	172	785	401
(5)	9	48	64	294	435

The estimated values, odds ratios, and log odds ratios are printed below. The estimated odds ratios and log odds ratios for the diagonal cells are all with a value of 2.89 and 1.06, respectively.

$$
\hat{m}_{ij} = \begin{bmatrix}
22.44 & 41.82 & 19.38 & 30.64 & 15.72 \\
33.23 & 179.17 & 83.01 & 131.27 & 67.32 \\
16.75 & 90.35 & 121.12 & 191.54 & 98.24 \\
23.70 & 127.81 & 171.34 & 784.03 & 402.11 \\
8.87 & 47.85 & 64.15 & 293.52 & 435.61
\end{bmatrix}
$$

$$
\hat{\theta}_{ij} = \begin{bmatrix}
2.89 & 1 & 1 & 1 \\
1 & 2.89 & 1 & 1 \\
1 & 1 & 2.89 & 1 \\
1 & 1 & 1 & 2.89
\end{bmatrix}
$$

$$
\hat{\Phi}_{ij} = \begin{bmatrix}
1.06 & 0 & 0 & 0 \\
0 & 1.06 & 0 & 0 \\
0 & 0 & 1.06 & 0 \\
0 & 0 & 0 & 1.06
\end{bmatrix}
$$

6.2.2   Variable Distance Model

The variable distance model (V), (Goodman 1972; Haberman 1974), has the multiplicative form:

$$
f_{ij} = \begin{cases}
\prod_{k=i}^{j-1} \delta_k & i < j \\
\prod_{k=j}^{i-1} \delta_k & i > j
\end{cases}
$$

and where $\delta_1, \delta_2, \ldots, \delta_I$ are the distances from categories 1 to 2, 2 to 3,..., and $(I - 1)$ to I, respectively. It assumes intervals are different among the categories. The main characteristic of variable distance model is that the estimated odds ratios are all one with the exception of the diagonal cells. While the diagonal cells of fixed distance model have a common odds ratio, the diagonal cells of variable distance model vary in their odds ratios (Fig. 6.3).

Fig. 6.3 Variable distance model

	1	2	3	4
1	θ_{11}	1	1	1
2	1	θ_{22}	1	1
3	1	1	θ_{33}	1
4	1	1	1	θ_{44}

For variable distance model, the nonstandard log-linear model is rewritten below with factor variables V1, V2, V3, and V4 specified for a 5 × 5 doubly classified table.

$$\ell_{ij} = \mu + \lambda_i^A + \lambda_j^B + \varphi_1 V1_{ij} + \varphi_1 V2_{2ij} + \varphi_3 V3_{ij} + \varphi_4 V4_{ij}$$

The variables V1, V2, V3, and V4 are specified below. It is noted that these 4 variables can be specified as numeric variables.

$$V1 = \begin{bmatrix} 2 & 1 & 1 & 1 & 1 \\ 1 & 2 & 2 & 2 & 2 \\ 1 & 2 & 2 & 2 & 2 \\ 1 & 2 & 2 & 2 & 2 \\ 1 & 2 & 2 & 2 & 2 \end{bmatrix} \quad V2 = \begin{bmatrix} 3 & 3 & 1 & 1 & 1 \\ 3 & 3 & 1 & 1 & 1 \\ 1 & 1 & 3 & 3 & 3 \\ 1 & 1 & 3 & 3 & 3 \\ 1 & 1 & 3 & 3 & 3 \end{bmatrix}$$

$$V3 = \begin{bmatrix} 4 & 4 & 4 & 1 & 1 \\ 4 & 4 & 4 & 1 & 1 \\ 4 & 4 & 4 & 1 & 1 \\ 1 & 1 & 1 & 4 & 4 \\ 1 & 1 & 1 & 4 & 4 \end{bmatrix} \quad V4 = \begin{bmatrix} 5 & 5 & 5 & 5 & 1 \\ 5 & 5 & 5 & 5 & 1 \\ 5 & 5 & 5 & 5 & 1 \\ 5 & 5 & 5 & 5 & 1 \\ 1 & 1 & 1 & 1 & 5 \end{bmatrix}$$

Table 6.4 displays the occupation status of father and son.
The R program for variable distance model for the above table is listed below.

Table 6.4 Occupation status of father and son

Father's status	Son's status				
	(1)	(2)	(3)	(4)	(5)
(1)	50	45	8	18	8
(2)	28	174	84	154	55
(3)	11	78	110	223	96
(4)	14	150	185	714	447
(5)	3	42	72	320	411

```
SonFather <- c (50,  45,  8,   18,   8,
               28, 174, 84, 154,  55,
               11,  78, 110, 223,  96,
               14, 150, 185, 714, 447,
                3,  42,  72, 320, 411)
r <- c(1, 1, 1, 1, 1,
       2, 2, 2, 2, 2,
       3, 3, 3, 3, 3,
       4, 4, 4, 4, 4,
       5, 5, 5, 5, 5)
r <- as.factorI
c <- c(1, 2, 3, 4, 5,
       1, 2, 3, 4, 5,
       1, 2, 3, 4, 5,
       1, 2, 3, 4, 5,
       1, 2, 3, 4, 5)
c <- as.factorI
V1 <- c(2, 1, 1, 1, 1,
        1, 2, 2, 2, 2,
        1, 2, 2, 2, 2,
        1, 2, 2, 2, 2,
        1, 2, 2, 2, 2)
V1 <- as.factor(V1)
V2 <- c(3, 3, 1, 1, 1,
        3, 3, 1, 1, 1,
        1, 1, 3, 3, 3,
        1, 1, 3, 3, 3,
        1, 1, 3, 3, 3)
V2 <- as.factor(V2)
V3 <- c(4, 4, 4, 1, 1,
        4, 4, 4, 1, 1,
        4, 4, 4, 1, 1,
        1, 1, 1, 4, 4,
        1, 1, 1, 4, 4)
V3 <- as.factor(V3)
V4 <- c(5, 5, 5, 5, 1,
        5, 5, 5, 5, 1,
        5, 5, 5, 5, 1,
        5, 5, 5, 5, 1,
        1, 1, 1, 1, 5)
V4 <- as.factor(V4)
VarDistance <- glm(SonFather ~ r + c + V1 + V2 + V3 + V4, family=poisson)
summary(VarDistance)
model.summary(VarDistance )
```

The estimated values and odds ratios are given below.

$$\hat{m}_{ij} = \begin{bmatrix} 50 & 32.08 & 12.46 & 22.40 & 12.07 \\ 23.46 & 191.46 & 74.35 & 133.69 & 72.04 \\ 9.91 & 80.88 & 113.40 & 203.92 & 109.88 \\ 16.18 & 132.03 & 185.12 & 764.65 & 412.01 \\ 6.44 & 52.55 & 73.68 & 304.33 & 411' \end{bmatrix}$$

$$\hat{\theta}_{ij} = \begin{bmatrix} 12.71 & 1 & 1 & 1 \\ 1 & 3.61 & 1 & 1 \\ 1 & 1 & 2.30 & 1 \\ 1 & 1 & 1 & 2.51 \end{bmatrix}$$

The following prints the estimated coefficients of the variable distance model.

```
> summary(VarDistance )

Call:
glm(formula = SonFather ~ r + c + V1 + V2 + V3 + V4, family = poisson)

Deviance Residuals:
    Min      1Q   Median      3Q     Max
-2.0955  -1.2817  -0.1962  0.9083  2.1498

Coefficients:
            Estimate Std. Error z value Pr(>|z|)
(Intercept) 1.12333    0.12406   9.055  < 2e-16 ***
r2          0.51501    0.11976   4.300 1.71e-05 ***
r3          0.29528    0.12121   2.436   0.0148 *
r4          1.20114    0.11573  10.379  < 2e-16 ***
r5          0.73926    0.11903   6.210 5.28e-10 ***
c2          0.82763    0.11983   6.907 4.96e-12 ***
c3          0.52364    0.12219   4.285 1.82e-05 ***
c4          1.52625    0.11596  13.162  < 2e-16 ***
c5          1.36731    0.11903  11.487  < 2e-16 ***
V12         1.27153    0.11506  11.051  < 2e-16 ***
V23         0.64194    0.05816  11.037  < 2e-16 ***
V34         0.41581    0.04563   9.114  < 2e-16 ***
V45         0.45942    0.04212  10.908  < 2e-16 ***
---
Signif. codes:  0 '***' 0.001 '**' 0.01 '*' 0.05 '.' 0.1 ' ' 1
```

The estimated values of φ_1, φ_2, φ_3, and φ_4 are 1.27153 (V12), 0.64194 (V23), 0.41581 (V34), and 0.45942 (V45) respectively. The local odds ratios can be derived from these estimates.

$$\Phi_{ij} = \begin{pmatrix} 2\varphi_1 & 0 & 0 & 0 \\ 0 & 2\varphi_2 & 0 & 0 \\ 0 & 0 & 2\varphi_3 & 0 \\ 0 & 0 & 0 & 2\varphi_4 \end{pmatrix} = \hat{\Phi}_{ij} = \begin{bmatrix} 2.54 & 0 & 0 & 0 \\ 0 & 1.28 & 0 & 0 \\ 0 & 0 & 0.83 & 0 \\ 0 & 0 & 0 & 0.92 \end{bmatrix}$$

$\hat{\Phi}_{11} = 2\delta_1 = 2 \times 1.27153 = 2.54$

$\hat{\Phi}_{22} = 2\delta_2 = 2 \times 0.64194 = 1.28$

$\hat{\Phi}_{33} = 2\delta_3 = 2 \times 0.41581 = 0.83$

$\hat{\Phi}_{44} = 2\delta_4 = 2 \times 0.45942 = 0.92$

6.3 Diagonal Band Models

Diagonal Band Model
Diagonal band models have the property that the estimated log odds ratios are equal to zero $(\Phi_{ij} = 0)$ unless $|i - j| \leq 1$. The diagonal band model is also known as null non-independence model (Goodman 1996).

$$\Phi_{ij} = 0 \text{ or } \theta_{ij} = 1 \text{ unless } |i - j| \leq 1$$
$$\Phi_{ij} = \ln(\theta_{ij})$$
$$\theta_{ij} : \text{odds ratio}$$

Models under diagonal band models include the uniform loyalty, the quasi-independence, and the triangle parameter models. The following subsections discuss these models in details.

6.3.1 Uniform Loyalty Model

Uniform loyalty model belongs to the group of diagonal band models. The odds ratios of the diagonal cells and those cells one position away from the diagonal are not equal to one, but the rest of the cells are equal to one. This model has also been referred to as the constant loyalty model, uniform loyalty model (Upton and Sarvik 1981), the uniform inheritance model (Goodman 1979a), and the smoothed-quasi-independence model (Scheuren and Loch 1975). The symbolic tables below describe the two main characteristic of uniform loyalty model. The left symbolic table shows that the diagonal odds ratios have the same value θ_1 and those cells that are one off-diagonal, their odds ratios are symmetrical and having the same value θ_2. The right symbolic table shows in log odds ratios. This is similar to that of the odds ratios representation but added that the association of Φ_1 and $\Phi_2 : \Phi_1 = -2\Phi_2$ (Fig. 6.4).

	1	2	3	4
1	θ_1	θ_2	1	1
2	θ_2	θ_1	θ_2	1
3	1	θ_2	θ_1	θ_2
4	1	1	θ_2	θ_1

	1	2	3	4
1	Φ_1	Φ_2	1	1
2	Φ_2	Φ_1	Φ_2	1
3	1	Φ_2	Φ_1	Φ_2
4	1	1	Φ_2	Φ_1

$$\Phi_1 = -2\Phi_2$$

Fig. 6.4 Uniform loyalty model

The nonstandard log-linear specification of the uniform loyalty model is as follows:

$$\ell_{ij} = \mu + \lambda_i^R + \lambda_j^C + \phi L_{ij}$$

$$L = \begin{cases} 1 & i \neq j \\ 2 & i = j \end{cases}$$

where L is a regression variable specified below for a 5×5 table.

$$L = \begin{pmatrix} 2 & 1 & 1 & 1 & 1 \\ 1 & 2 & 1 & 1 & 1 \\ 1 & 1 & 2 & 1 & 1 \\ 1 & 1 & 1 & 2 & 1 \\ 1 & 1 & 1 & 1 & 2 \end{pmatrix}$$

As the absolute of the local odds ratios of the diagonal are one half of those one position off-diagonal, the estimated structure of Φ_{ij} that can be derived from the estimated coefficient ϕ is shown below.

$$\Phi_{ij} = \begin{pmatrix} 2\phi & -\phi & 0 & 0 \\ -\phi & 2\phi & -\phi & 0 \\ 0 & -\phi_3 & 2\phi & -\phi \\ 0 & 0 & -\phi & 2\phi \end{pmatrix}$$

Uniform loyalty model is often used in mobility studies. In the language of mobility, this model differentiates those who do not change from those who do change. The model differentiates between the diagonal and the off-diagonal cells where the diagonal members are assumed to be homogeneous, i.e., they are all assumed to have the same probability of inheritance.

Table 6.5 tabulates the occupation status of father and son.

The R syntax to generate uniform loyalty model for the above table is as follows:

```
r <- c(1, 1, 1, 1, 1,
       2, 2, 2, 2, 2,
       3, 3, 3, 3, 3,
       4, 4, 4, 4, 4,
       5, 5, 5, 5, 5)
```

Table 6.5 Occupation status of father and son

Son's occupation status	Father's occupation status				
	1	2	3	4	5
1	9	22	20	50	40
2	16	130	65	160	130
3	21	70	135	170	137
4	40	170	162	800	330
5	27	100	95	245	389

```
r <- as.factorI
c <- c(1, 2, 3, 4, 5,
       1, 2, 3, 4, 5,
       1, 2, 3, 4, 5,
       1, 2, 3, 4, 5,
       1, 2, 3, 4, 5)
c <- as.factorI
# with and without specifying as factor gives same result
L  <- c(2, 1, 1, 1, 1,
        1, 2, 1, 1, 1,
        1, 1, 2, 1, 1,
        1, 1, 1, 2, 1,
        1, 1, 1, 1, 2)
L <- as.factor(L)
Uloyalty <- glm(BritishSonFather ~ r + c + L, family=poisson)
summary(Uloyalty)
model.summary(Uloyalty)
```

The following prints the estimated coefficients of uniform loyalty model.

```
> summary(ULoyalty)

Call:
glm(formula = URM ~ r + c + L, family = poisson)

Deviance Residuals:
     Min        1Q    Median        3Q       Max
 -0.40112  -0.11255   0.00945   0.06795   0.70838

Coefficients:
             Estimate Std. Error z value Pr(>|z|)
(Intercept)   0.97549    0.12576   7.757 8.7e-15 ***
r2            1.16716    0.09581  12.182 < 2e-16 ***
r3            1.23410    0.09515  12.970 < 2e-16 ***
r4            2.09881    0.08965  23.411 < 2e-16 ***
r5            1.58771    0.09221  17.219 < 2e-16 ***
c2            1.37694    0.10474  13.146 < 2e-16 ***
c3            1.33699    0.10509  12.722 < 2e-16 ***
c4            2.25745    0.09926  22.742 < 2e-16 ***
c5            2.04590    0.09984  20.493 < 2e-16 ***
L             0.67627    0.03660  18.475 < 2e-16 ***
---
Signif. codes:  0 '***' 0.001 '**' 0.01 '*' 0.05 '.' 0.1 ' ' 1
```

The estimated values, odds ratios, and log odds ratios are shown below. The estimated coefficient L (0.67627) is the negative value of log odds ratios for the one off-diagonal cell, $\hat{\phi} = 0.68$ and $\hat{\Phi}_{ii} = -0.68$.

$$\hat{m}_{ij} = \begin{bmatrix} 10.26 & 20.67 & 19.86 & 49.86 & 40.35 \\ 16.76 & 130.60 & 63.81 & 160.19 & 129.65 \\ 17.92 & 71.01 & 134.17 & 171.28 & 138.62 \\ 42.55 & 168.60 & 161.99 & 799.73 & 329.13 \\ 25.52 & 101.13 & 97.17 & 243.94 & 388.24 \end{bmatrix}$$

$$\hat{\theta}_{ij} = \begin{bmatrix} 3.87 & 0.51 & 1 & 1 \\ 0.51 & 3.87 & 0.51 & 1 \\ 1 & 0.51 & 3.87 & 0.51 \\ 1 & 1 & 0.51 & 3.87 \end{bmatrix}$$

$$\hat{\Phi}_{ij} = \begin{pmatrix} 2\hat{\phi} & -\hat{\phi} & 0 & 0 \\ -\hat{\phi} & 2\hat{\phi} & -\hat{\phi} & 0 \\ 0 & -\hat{\phi} & 2\hat{\phi} & -\hat{\phi} \\ 0 & 0 & -\hat{\phi} & 2\hat{\phi} \end{pmatrix} = \begin{bmatrix} 1.35 & -0.68 & 0 & 0 \\ -0.68 & 1.35 & -0.68 & 0 \\ 0 & -0.68 & 1.35 & -0.68 \\ 0 & 0 & -0.68 & 1.35 \end{bmatrix}$$

6.3.2 Quasi Independence Model

The quasi independence (QI) model is yet another diagonal band model, a generalization of the independence model. The model of quasi independence (QI) has the following multiplicative form:

$$m_{ij} = \alpha_i \beta_j \psi_{ij}$$

where α_i and β_j denote parameters pertaining to the ith row and the jth column, respectively, of an $a \times a$ table. QI has been described as variable loyalty model (Upton and Sarvik 1981), non-uniform loyalty model (Scheuren and Loch 1975), and mover–stayer model (Upton and Sarvik 1981). The model has one more parameter Q than the independence model and is therefore based on $(a^2 - 3a + 1)$ degrees of freedom. The main characteristics of quasi independence model are that the odds ratios of the diagonal have different values, whereas the one off-diagonal odds ratio is symmetrical, as shown in the symbolic table (Fig. 6.5).

Fig. 6.5 Quasi independent model

	1	2	3	4
1	θ_{11}	θ_2	1	1
2	θ_2	θ_{22}	θ_3	1
3	1	θ_3	θ_{33}	θ_4
4	1	1	θ_4	θ_{44}

The following is the corresponding nonstandard log-linear model formulation of quasi independence model.

$$\ell_{ij} = \mu + \lambda_i^R + \lambda_j^C + \lambda_{ij}^Q$$

where Q is a factor variable as shown below.

$$Q = \begin{cases} 1 & i \neq j \\ i+1 & i = j \end{cases}$$

For a 5×5 table, the Q factor is specified as follows:

$$Q = \begin{pmatrix} 2 & 1 & 1 & 1 & 1 \\ 1 & 3 & 1 & 1 & 1 \\ 1 & 1 & 4 & 1 & 1 \\ 1 & 1 & 1 & 5 & 1 \\ 1 & 1 & 1 & 1 & 6 \end{pmatrix}$$

The following cross-tabulation tabulates the occupation status of father and son (Table 6.6).

The R syntax to generate quasi independence model for the above table is shown below.

```
QIM <- c(50, 15, 10, 40, 20,
          7, 170, 50, 175, 90,
          9, 60, 110, 230, 120,
          30, 180, 210, 715, 380,
          15, 70, 80, 280, 420)
r <- c(1, 1, 1, 1, 1,
        2, 2, 2, 2, 2,
        3, 3, 3, 3, 3,
        4, 4, 4, 4, 4,
        5, 5, 5, 5, 5)
r <- as.factor
c <- c(1, 2, 3, 4, 5,
        1, 2, 3, 4, 5,
```

Table 6.6 Father's and son's occupation status

Son's occupation status	Father's occupation status				
	1	2	3	4	5
1	50	15	10	40	20
2	7	170	50	175	90
3	9	60	110	230	120
4	30	180	210	715	380
5	15	70	80	280	420

```
        1, 2, 3, 4, 5,
        1, 2, 3, 4, 5,
        1, 2, 3, 4, 5)
c <- as.factor
Q <- c(2, 1, 1, 1, 1,
       1, 3, 1, 1, 1,
       1, 1, 4, 1, 1,
       1, 1, 1, 5, 1,
       1, 1, 1, 1, 6)
Q <- as.factor(Q)
QuasiInd <- glm(QIM ~ r + c + Q, family=poisson)
summary(QuasiInd)
model.summary(QuasiInd)
```

Factor *Q* can be generated using package catspec function mob.qi(). The syntax to generate it is stated below. The function fitmacro() generates the fit statistics.

```
library(catspec)
QuasiDF <- as.data.frame(cbind(r,c,QIM))
QuasiDF$r <- as.factor(QuasiDF$r)
QuasiDF$c <- as.factor(QuasiDF$c)
glm.QI<-glm(QIM ~ r + c + mob.qi(r,c),family=poisson,data=QuasiDF)
summary(glm.QI)
fitmacro(glm.QI)
```

The output from function glm.QI() is printed below.

```
> fitmacro(glm.QI)

deviance:                3.178
df:                         11
bic:                    -86.700
aic:                    -18.822
Number of parameters:       14
Number of cases:          3536
```

The estimated values, odds ratios, and log odds ratios are printed below.

$$\hat{m}_{ij} = \begin{bmatrix} 50 & 10.51 & 11.84 & 41.12 & 21.53 \\ 7.54 & 170 & 49.98 & 173.58 & 90.89 \\ 9.99 & 58.75 & 110 & 229.89 & 120.38 \\ 31.29 & 184.08 & 207.43 & 715 & 377.20 \\ 12.18 & 71.66 & 80.75 & 280.41 & 420 \end{bmatrix}$$

$$\hat{\theta}_{ij} = \begin{bmatrix} 107.26 & 0.26 & 1 & 1 \\ 0.26 & 6.37 & 0.60 & 1 \\ 1 & 0.60 & 1.65 & 1.01 \\ 1 & 1 & 1.01 & 2.84 \end{bmatrix}$$

$$\hat{\Phi}_{ij} = \begin{bmatrix} 4.68 & -1.34 & 0 & 0 \\ -1.34 & 1.85 & -0.51 & 0 \\ 0 & -0.51 & 0.50 & 0.01 \\ 0 & 0 & 0.01 & 1.04 \end{bmatrix}$$

6.3.3 Triangle Parameters Model

Triangle parameters model has the following multiplicative form (Goodman 1972). The main characteristic of triangle parameters model is there are three common odds ratios. One for the diagonal cells, θ_1, one for the one off-diagonal at upper diagonal, θ_2, and the third for the lower diagonal, θ_3. The symbolic table below shows these features. The estimated log odds ratios of triangle parameters for the three log odds ratios are also stated below where $\hat{\tau}_1$ and $\hat{\tau}_2$ pertain to the upper right and lower left one off-diagonal estimate of log odds ratios, and $-(\hat{\tau}_1 + \hat{\tau}_2)$ is the estimated diagonal log odds ratio (Fig. 6.6).

$$m_{ij} = \alpha_i \beta_j \gamma_{ij}$$

where

Fig. 6.6 Triangle parameters model

	1	2	3	4
1	θ_1	θ_2	1	1
2	θ_3	θ_1	θ_2	1
3	1	θ_3	θ_1	θ_2
4	1	1	θ_3	θ_1

$$\gamma_{ij} = \begin{cases} \tau_1 & i > j \\ \tau_2 & i < j \\ 1 & i = j \end{cases}, \quad \hat{\Phi}_{ij} = \begin{cases} \hat{\tau}_1 & i > j \\ -(\hat{\tau}_1 + \hat{\tau}_2) & i = j \\ \hat{\tau}_2 & i < j \end{cases}$$

The nonstandard log-linear specification of triangle parameters model is stated below.

$$\ell_{ij} = \mu + \lambda_i^R + \lambda_j^C + \lambda_{ij}^T$$

where T is a factor variable as shown below.

$$T = \begin{cases} 1 & i = j \\ 2 & i > j \\ 3 & j > i \end{cases}$$

For a 5×5 table, the T factor is specified as follows:

$$T = \begin{pmatrix} 1 & 3 & 3 & 3 & 3 \\ 2 & 1 & 3 & 3 & 3 \\ 2 & 2 & 1 & 3 & 3 \\ 2 & 2 & 2 & 1 & 3 \\ 2 & 2 & 2 & 2 & 1 \end{pmatrix}$$

Table 6.7 tabulates the occupation status of son and father.
The R syntax to generate the triangle parameters model is stated below.

```
Triangle <- c(10, 21, 19, 50, 40,
              20, 135, 65, 160, 125,
              17, 72, 130, 178, 137,
              45, 170, 150, 800, 340,
              30, 107, 100, 250, 400)
r <- c(1, 1, 1, 1, 1,
       2, 2, 2, 2, 2,
       3, 3, 3, 3, 3,
       4, 4, 4, 4, 4,
       5, 5, 5, 5, 5)
```

Table 6.7 Son's and father's occupation status	Son's occupation status	Father's occupation status				
		1	2	3	4	5
	1	10	21	19	50	40
	2	20	135	65	160	125
	3	17	72	130	178	137
	4	45	170	150	800	340
	5	30	107	100	250	400

```
r <- as.factorI
c <- c(1, 2, 3, 4, 5,
       1, 2, 3, 4, 5,
       1, 2, 3, 4, 5,
       1, 2, 3, 4, 5,
       1, 2, 3, 4, 5)
c <- as.factorI
T <- c(3, 1, 1, 1, 1,
       2, 3, 1, 1, 1,
       2, 2, 3, 1, 1,
       2, 2, 2, 3, 1,
       2, 2, 2, 2, 3)
T <- as.factor(T)
NonIndepTri <- glm(Triangle ~ r + c + T, family=poisson)
summary(NonIndepTri)
model.summary(NonIndepTri)
```

The estimated coefficients of $T2$ and $T3$ are -0.72281 and -0.62183, respectively, representing $\hat{\tau}_1$ and $\hat{\tau}_2$. The diagonal log odds ratio is $-(\hat{\tau}_1 + \hat{\tau}_2) = -(-0.62 - 0.72) = 1.34$.

```
Coefficients:
            Estimate Std. Error z value Pr(>|z|)
(Intercept)  2.38224    0.12773  18.651   <2e-16 ***
r2           1.19325    0.09654  12.360   <2e-16 ***
r3           1.27575    0.09895  12.892   <2e-16 ***
r4           2.16289    0.10492  20.614   <2e-16 ***
r5           1.71862    0.12844  13.381   <2e-16 ***
c2           1.31658    0.10191  12.919   <2e-16 ***
c3           1.20693    0.10600  11.387   <2e-16 ***
c4           2.13934    0.11168  19.157   <2e-16 ***
c5           1.89409    0.13297  14.245   <2e-16 ***
T2          -0.72291    0.06566 -11.010   <2e-16 ***
T3          -0.62183    0.06402  -9.713   <2e-16 ***
```

$$\hat{\Phi}_{ij} = \begin{bmatrix} -(\hat{\tau}_1 + \hat{\tau}_2) & \tau_2 & 0 & 0 \\ \tau_1 & -(\hat{\tau}_1 + \hat{\tau}_2) & \tau_2 & 0 \\ 0 & \tau_1 & -(\hat{\tau}_1 + \hat{\tau}_2) & \tau_2 \\ 0 & 0 & \tau_1 & -(\hat{\tau}_1 + \hat{\tau}_2) \end{bmatrix}$$

$$= \begin{bmatrix} 1.34 & -0.62 & 0 & 0 \\ -0.72 & 1.34 & -0.62 & 0 \\ 0 & -0.72 & 1.34 & -0.62 \\ 0 & 0 & -0.72 & 1.34 \end{bmatrix}$$

The estimated values and odds ratios are printed below.

$$\hat{m}_{ij} = \begin{bmatrix} 10.83 & 21.69 & 19.44 & 49.39 & 38.65 \\ 17.33 & 133.23 & 64.11 & 162.88 & 127.45 \\ 18.82 & 70.22 & 129.66 & 176.88 & 138.41 \\ 45.71 & 170.51 & 152.80 & 799.89 & 336.09 \\ 29.31 & 109.35 & 97.99 & 248.96 & 401.39 \end{bmatrix}$$

$$\hat{\theta}_{ij} = \begin{bmatrix} 3.84 & 0.54 & 1 & 1 \\ 0.49 & 3.84 & 0.54 & 1 \\ 1 & 0.49 & 3.84 & 0.54 \\ 1 & 1 & 0.49 & 3.84 \end{bmatrix}$$

6.4   Full Diagonal Models

This section covers full diagonal models. As mentioned earlier, models under it do not restrict the odds ratios on diagonal or those with distance away from the off-diagonal. Five models will be discussed in this section which include full diagonal model, diagonal absolute model, uniform association model, uniform fixed distance association model, and uniform variable distance association model. The last two models are considered as composite models as each of the two models combine the features of two models into a model.

6.4.1   Uniform Association Model

The first full diagonal model introduces in this section is uniform association model (U). It is also referred to as uniform diagonal model. This model has a main property that the local odds ratios are with the same value; that is, $\theta_{ij} = \theta$ for all the cell (i, j). The symbolic table below shows this main property of uniform association model. Uniform association model posits that the local association is constant throughout the entire table since all local odds ratios having the same value. The expected value of uniform association model is defined as follows:

$$m_{ij} = \alpha_i \beta_j \theta^{ij} \quad i, j = 1, 2, \ldots, a$$

where α_i and β_j are parameters relate to the marginal densities of the ith row and the jth column (Lawal 1992). It has $a(a-2)$ degrees of freedom (Fig. 6.7).

The nonstandard log-linear model for uniform association model is stated below. It is noted that U is not a factor variable but a quantitative variable.

Fig. 6.7 Uniform association
model

	1	2	3	4
1	θ	θ	θ	θ
2	θ	θ	θ	θ
3	θ	θ	θ	θ
4	θ	θ	θ	θ

$$\ell_{ij} = \mu + \lambda_i^R + \lambda_j^C + \kappa U_{ij}$$

The quantitative variable U for a 5×5 doubly classified table is shown below.
The first row is a running number from 1 to 5. The second row is a multiplication of
2 starting with 2, the third row is a multiplication of 3 starting from 3, the fourth is a
multiplication of 4 starting from 4, and so on.

$$U = \begin{pmatrix} 1 & 2 & 3 & 4 & 5 \\ 2 & 4 & 6 & 8 & 10 \\ 3 & 6 & 9 & 12 & 15 \\ 4 & 8 & 12 & 16 & 20 \\ 5 & 10 & 15 & 20 & 25 \end{pmatrix}$$

Table 6.8 tabulates occupation status of 3542 pairs of fathers and sons.
Occupation status is grouped into five categories being 1 representing the lowest
occupational status and 5 representing the highest occupational status.

The R syntax to generate uniform association model for the above table is listed
below.

```
Mobility <- c(42, 32, 25, 20, 15,
              95, 98, 97, 98, 98,
              56, 75, 90, 120, 175,
              79, 132, 240, 410, 670,
              20, 40, 95, 220, 482)
r <- c(1, 1, 1, 1, 1,
       2, 2, 2, 2, 2,
       3, 3, 3, 3, 3,
```

Table 6.8 Son's and father's
occupation status

Son's occupation status	Father's occupation status				
	1	2	3	4	5
1	42	32	25	20	15
2	95	98	97	98	98
3	56	75	90	120	175
4	79	132	240	410	670
5	20	40	95	220	482

```
        4, 4, 4, 4, 4,
        5, 5, 5, 5, 5)
r <- as.factor(r)
c <- c(1, 2, 3, 4, 5,
        1, 2, 3, 4, 5,
        1, 2, 3, 4, 5,
        1, 2, 3, 4, 5,
        1, 2, 3, 4, 5)
U <- c(1,  2,  3,  4,  5,
        2,  4,  6,  8, 10,
        3,  6,  9, 12, 15,
        4,  8, 12, 16, 20,
        5, 10, 15, 20, 25)
UAM <- glm(Mobility~ r + c + U, family=poisson)
summary(UAM)
model.summary(UAM)
```

The model estimates for the uniform association model are printed below.

```
Coefficients:
             Estimate Std. Error z value Pr(>|z|)
(Intercept)   3.47489   0.10005  34.732  < 2e-16 ***
r2            0.55620   0.10096   5.509 3.61e-08 ***
r3           -0.25505   0.11619  -2.195   0.0282 *
r4           -0.16539   0.14521  -1.139   0.2547
r5           -1.84318   0.19566  -9.420  < 2e-16 ***
c2           -0.53447   0.08417  -6.350 2.16e-10 ***
c3           -1.04881   0.10237 -10.245  < 2e-16 ***
c4           -1.56028   0.13846 -11.269  < 2e-16 ***
c5           -2.09944   0.18689 -11.234  < 2e-16 ***
U             0.26570   0.01324  20.069  < 2e-16 ***
```

The estimated frequencies and odds ratios are printed below. The estimated odds ratio can derived either from the estimated frequencies or from the estimated coefficient of U, $\exp(\hat{\kappa}) = \exp(0.2657) = 1.304$.

$$
\hat{m}_{ij} = \begin{bmatrix}
41.92 & 32.45 & 25.12 & 19.45 & 15.05 \\
95.47 & 96.33 & 97.19 & 98.07 & 98.95 \\
55.42 & 72.88 & 95.85 & 126.06 & 165.79 \\
79.22 & 135.80 & 232.80 & 399.07 & 684.10 \\
19.34 & 43.22 & 96.57 & 215.77 & 482.11
\end{bmatrix},
$$

$$
\hat{\theta}_{ij} = \begin{bmatrix}
1.304 & 1.304 & 1.304 & 1.304 \\
1.304 & 1.304 & 1.304 & 1.304 \\
1.304 & 1.304 & 1.304 & 1.304 \\
1.304 & 1.304 & 1.304 & 1.304
\end{bmatrix}
$$

The estimated local odds ratios are with the same value of 1.303. This means that the odds of those sons in a status n and those in status $n + 1$ that their fathers in status n resulted in the same odds ratio of 1.303. For instance, the odds of sons in status 1 and in status 2 that their fathers remain at status 1 are with an odds ratio of 1.303. Similarly, the odds of sons in status 2 and in status 3 that their fathers remain at status 1 are also with an odds ratio of 1.303. If local odds ratios are all equal to one, uniform association reduces to independence model. Thus, θ determines the level of association of relationship of cell.

6.4.2   Diagonal D Model

The diagonal D model is stated below. The symbolic table below shows the main characteristics of this model. The odds ratio of the main diagonal cells has one common odds ratio. The off-diagonal cells each have its own odds ratio (Fig. 6.8).

$$m_{ij} = \alpha_i \beta_j \delta_{i-j} = \alpha_i \beta_j \delta_k \quad \text{for } i \neq j$$

$$\text{where} \quad k = i - j$$

$$\theta_{ij} = \frac{(\alpha_i \beta_j \delta_{i-j})(\alpha_{i+1}\beta_{j+1}\delta_{i-j})}{(\alpha_{i+1}\beta_j \delta_{i-j+1})(\alpha_i \beta_{j+1}\delta_{i-j-1})} = \frac{\delta_k^2}{\delta_{k+1}\delta_{k-1}}$$

The nonstandard log-linear model for diagonal D model is stated below, with degrees of freedom $(a-2)^2$.

$$\ell_{ij} = \mu + \lambda_i^R + \lambda_j^C + \lambda_{ij}^D$$

$$D = \begin{pmatrix} 0 & -1 & -2 & -3 & -4 \\ 1 & 0 & -1 & -2 & -3 \\ 2 & 1 & 0 & -1 & -2 \\ 3 & 2 & 1 & 0 & -1 \\ 4 & 3 & 2 & 1 & 0 \end{pmatrix} \quad \text{or} \quad D = \begin{pmatrix} 1 & 6 & 7 & 8 & 9 \\ 2 & 1 & 6 & 7 & 8 \\ 3 & 2 & 1 & 6 & 7 \\ 4 & 3 & 2 & 1 & 6 \\ 5 & 4 & 3 & 2 & 1 \end{pmatrix}$$

where D is the 5×5 factor variable specified above. This definition of the factor variable takes into account the appropriate values of $k = (i - j)$. The above gives two different form of the factor variable D that leads to the same model fit. However, the parameter estimates differ.

Fig. 6.8 Diagonal D model

	1	2	3	4
1	θ_{11}	θ_{12}	θ_{13}	θ_{14}
2	θ_{21}	θ_{11}	θ_{23}	θ_{24}
3	θ_{31}	θ_{32}	θ_{11}	θ_{34}
4	θ_{41}	θ_{42}	θ_{43}	θ_{11}

Table 6.9 Father's and son's occupation status

Son's occupation status	Father's occupation status				
	1	2	3	4	5
1	25	50	20	25	9
2	40	175	100	140	50
3	20	90	105	210	96
4	20	146	168	737	445
5	5	37	72	325	420

Table 6.9 shows the tabulation of 3530 pairs of father's and son's occupation status.

The R syntax for generating diagonal D model for the above table is stated below.

```
DiagD <- c(25,  50,  20,  25,   9,
           40, 175, 100, 140,  50,
           20,  90, 105, 210,  96,
           20, 146, 168, 737, 445,
            5,  37,  72, 325, 420)
D <- c(1, 6, 7, 8, 9,
       2, 1, 6, 7, 8,
       3, 2, 1, 6, 7,
       4, 3, 2, 1, 6,
       5, 4, 3, 2, 1)
D <- as.factor(D)
r <- c(1, 1, 1, 1, 1,
       2, 2, 2, 2, 2,
       3, 3, 3, 3, 3,
       4, 4, 4, 4, 4,
       5, 5, 5, 5, 5)
r <- as.factorI
c <- c(1, 2, 3, 4, 5,
       1, 2, 3, 4, 5,
       1, 2, 3, 4, 5,
       1, 2, 3, 4, 5,
       1, 2, 3, 4, 5)
c <- as.factorI
DiagDM <- glm(DiagD ~ r + c + D, family=poisson)
summary(DiagDM)
model.summary(DiagDM)
```

The fitted values and estimated odds ratios are printed below. The parallel in odds ratios is obviously noted. For instance, $\hat{\theta}_{11} = \hat{\theta}_{22} = \hat{\theta}_{33} = \hat{\theta}_{44} = 2.14$, $\hat{\theta}_{12} = \hat{\theta}_{23} = \hat{\theta}_{34} = 1.37$, and so on.

$$\hat{m}_{ij} = \begin{bmatrix} 24.88 & 50.73 & 20.23 & 24.16 & 9.00 \\ 40.27 & 176.11 & 96.00 & 141.78 & 50.84 \\ 20.32 & 89.79 & 104.97 & 211.94 & 93.99 \\ 19.53 & 143.90 & 170.02 & 736.21 & 446.33 \\ 5.00 & 37.47 & 73.78 & 322.91 & 419.84 \end{bmatrix}$$

$$\hat{\theta}_{ij} = \begin{bmatrix} 2.14 & 1.37 & 1.24 & 0.96 \\ 1.01 & 2.14 & 1.37 & 1.24 \\ 1.67 & 1.01 & 2.14 & 1.37 \\ 1.02 & 1.67 & 1.01 & 2.14 \end{bmatrix}$$

6.4.3 Diagonal Absolute Model

The third model under the full diagonal models discusses in this section is diagonal absolute model (DA). The symbolic table below shows the main characteristics of this model. All odds ratios are symmetrical from the diagonal. Those cells that belonged to the same position away from diagonal are with the same value. For instance, the odds ratios of cell (1,3), cell (2,4), cell (3,1), and cell (4,2) are all with a constant θ_3. The constant odds ratio also applies to the diagonal cells having the same odds ratio θ_1, i.e., the odds ratios of cell (1,1), cell (2,2), cell (3,3), and cell (4,4) are all with a constant θ_1. Diagonal absolute model has the following multiplicative form with $(a-1)(a-2)$ degrees of freedom (Fig. 6.9).

$$m_{ij} = \alpha_i \beta_j \delta_k \quad \text{for } i \neq j$$

where $k = |i - j|$.

The nonstandard log-linear specification of diagonal absolute model is stated below.

Fig. 6.9 Diagonal absolute model

	1	2	3	4
1	θ_1	θ_2	θ_3	θ_4
2	θ_2	θ_1	θ_2	θ_3
3	θ_3	θ_2	θ_1	θ_2
4	θ_4	θ_3	θ_2	θ_1

$$\ell_{ij} = \mu + \lambda_i^R + \lambda_j^C + \lambda_{ij}^{DA}$$

$$\text{where DA} = \begin{cases} 1 & i = j \\ k+1 & |i-k| = k \end{cases}$$

For 5 × 5 table, the factor DA is specified as below.

$$DA = \begin{pmatrix} 1 & 2 & 3 & 4 & 5 \\ 2 & 1 & 2 & 3 & 4 \\ 3 & 2 & 1 & 2 & 3 \\ 4 & 3 & 2 & 1 & 2 \\ 5 & 4 & 3 & 2 & 1 \end{pmatrix}$$

The frequency tabulation of son's and father's occupation status is tabulated in Table 6.10.

The R syntax to generate the diagonal absolute model for the above table is stated below.

```
DiagAbsData <- c(25, 50, 22, 25, 8,
                 40, 170, 89, 145, 50,
                 20, 95, 104, 200, 100,
                 20, 140, 180, 740, 440,
                 5, 35, 70, 325, 420)
r <- c(1, 1, 1, 1, 1,
       2, 2, 2, 2, 2,
       3, 3, 3, 3, 3,
       4, 4, 4, 4, 4,
       5, 5, 5, 5, 5)
r <- as.factorI
c <- c(1, 2, 3, 4, 5,
       1, 2, 3, 4, 5,
       1, 2, 3, 4, 5,
       1, 2, 3, 4, 5,
       1, 2, 3, 4, 5)
c <- as.factorI
DA <- c(1, 2, 3, 4, 5,
```

Table 6.10 Son's and father's occupation status

Son's occupation status	Father's occupation status				
	1	2	3	4	5
1	25	50	22	25	8
2	40	170	89	145	50
3	20	95	104	200	100
4	20	140	180	740	440
5	5	35	70	325	420

```
     2, 1, 2, 3, 4,
     3, 2, 1, 2, 3,
     4, 3, 2, 1, 2,
     5, 4, 3, 2, 1)
DA <- as.factor(DA)
DiagAbs <- glm(DiagAbsData~ r + c + DA, family=poisson)
summary(DiagAbs)
model.summary(DiagAbs)
```

The fitted values, odds ratios, and local odds ratios are printed below. It is observed that the estimated odds ratios of the diagonal are the same with value 2.15. Those odds ratios that are one position away from diagonal are with a value of 1.16, two positions away are with a constant value of 1.48, and three positions are with a value 1.04. The same can be observed from the log odds ratios.

$$\hat{m}_{ij} = \begin{bmatrix} 25.55 & 49.62 & 22.16 & 24.50 & 8.17 \\ 40.75 & 170.44 & 88.23 & `144.57 & 50 \\ 19.38 & 93.96 & 104.75 & 198.96 & 101.96 \\ 19.49 & 139.98 & 180.91 & 740.01 & 439.61 \\ 4.83 & 36.01 & 68.95 & 326.95 & 418.26 \end{bmatrix}$$

$$\hat{\theta}_{ij} = \begin{bmatrix} 2.15 & 1.16 & 1.48 & 1.04 \\ 1.16 & 2.15 & 1.16 & 1.48 \\ 1.48 & 1.16 & 2.15 & 1.16 \\ 1.04 & 1.48 & 1.16 & 2.15 \end{bmatrix}$$

$$\hat{\Phi}_{ij} = \begin{bmatrix} 0.77 & 0.15 & 0.39 & 0.04 \\ 0.15 & 0.77 & 0.15 & 0.39 \\ 0.39 & 0.15 & 0.77 & 0.15 \\ 0.04 & 0.39 & 0.15 & 0.77 \end{bmatrix}$$

Relationships of Independence, Diagonal Absolute, and Uniform Association and Fixed Distance Model

The relationships of the two full diagonal models (uniform association and diagonal absolute models) so far discussed, and the independence model and fixed distance model are depicted in the diagram below. If all local odds ratios are equal to one, uniform association model reduces to independence model. If all the diagonal θs reduce to one, fixed distance model becomes independence model. Similarly, if all the parallel diagonal θs reduce to one, diagonal absolute model becomes independence model. Uniform association model reduces to independence model if all the θs reduce to one (Fig. 6.10).

Fig. 6.10 Relationships of independence, diagonal absolute, and uniform association and fixed distance model

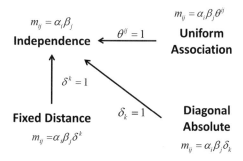

$$m_{ij} = \alpha_i \beta_j$$
Independence $\xleftarrow{\theta^{ij} = 1}$

$$m_{ij} = \alpha_i \beta_j \theta^{ij}$$
Uniform Association

$\delta^k = 1$

Fixed Distance $\quad \delta_k = 1 \searrow \quad$ **Diagonal Absolute**

$$m_{ij} = \alpha_i \beta_j \delta^k$$

$$m_{ij} = \alpha_i \beta_j \delta_k$$

6.4.4 Uniform Fixed Distance Association Model

The various non-independence models so far discussed may not meet the complex variations encounter in real-life application. Second-order model becomes an option. It combines models. The uniform association model and fixed distance model can be combined into a model. Lawal (1992) called this model as uniform fixed distance association model (UF). The expected frequencies under this model are stated below. There are two additional parameters δ and θ more than the independent model. It has $(a^2 - 2a - 1)$ degrees of freedom.

$$m_{ij} = \alpha_i \beta_j \delta_k \theta^{ij} \quad \text{for } k = |i - j|$$

The symbolic table below shows the characteristic of uniform fixed distance association model. The main diagonal has one common odds ratio θ_1, and the off-diagonal contains another common odds ratio θ_2 (Fig. 6.11).

The nonstandard log-linear model for uniform association is stated below. It is noted that U is not a factor variable but a quantitative variable.

$$\ell_{ij} = \mu + \lambda_i^R + \lambda_j^C + \lambda_{ij}^U + \varphi F_{ij}$$

The quantitative variables U, as previously defined, are equal to $i \times j$ for cell (i, j) as shown below for a 5×5 square model. The F regression variable is also restated below.

Fig. 6.11 Uniform fixed distance association model

	1	2	3	4
1	θ_1	θ_2	θ_2	θ_2
2	θ_2	θ_1	θ_2	θ_2
3	θ_2	θ_2	θ_1	θ_2
4	θ_2	θ_2	θ_2	θ_1

$$U = \begin{pmatrix} 1 & 2 & 3 & 4 & 5 \\ 2 & 4 & 6 & 8 & 10 \\ 3 & 6 & 9 & 12 & 15 \\ 4 & 8 & 12 & 16 & 20 \\ 5 & 10 & 15 & 20 & 25 \end{pmatrix} \quad F = \begin{pmatrix} 1 & 2 & 3 & 4 & 5 \\ 1 & 1 & 2 & 3 & 4 \\ 1 & 1 & 1 & 2 & 3 \\ 1 & 1 & 1 & 1 & 2 \\ 1 & 1 & 1 & 1 & 1 \end{pmatrix}$$

Table 6.11 is a well-analyzed social mobility data about pair of Danish son's and father's occupation status.

The specification of the uniform association and fixed distance model is simply to place both the F and U factors into the model as shown in the R syntax below.

```
DanishSonFather <- c (18,  17,  16,   4,   2,
                      24, 105, 109,  59,  21,
                      23,  84, 289, 217,  95,
                       8,  49, 175, 384, 198,
                       6,   8,  69, 201, 246)
F <- c(1, 2, 3, 4, 5,
       1, 1, 2, 3, 4,
       1, 1, 1, 2, 3,
       1, 1, 1, 1, 2,
       1, 1, 1, 1, 1)
U <- c(1,  2,  3,  4,  5,
       2,  4,  6,  8, 10,
       3,  6,  9, 12, 15,
       4,  8, 12, 16, 20,
       5, 10, 15, 20, 25)
UniformFixed <- glm(DanishSonFather ~ r + c + F + U, family = poisson)
summary(UniformFixed)
model.summary(UniformFixed)
```

The estimated values and odds ratios are printed below.

Father's status	Son's status				
	1	2	3	4	5
1	18	17	16	4	2
2	24	105	109	59	21
3	23	84	289	217	95
4	8	49	175	384	198
5	6	8	69	201	246

Table 6.11 Occupation status of Danish father–son pairs

$$\hat{m}_{ij} = \begin{bmatrix} 15.13 & 17.19 & 15.35 & 7.46 & 1.87 \\ 29.89 & 95.33 & 106.95 & 65.23 & 20.60 \\ 22.75 & 91.15 & 287.02 & 219.86 & 87.22 \\ 9.04 & 45.49 & 179.93 & 386.81 & 192.73 \\ 2.19 & 13.84 & 68.75 & 185.64 & 259.58 \end{bmatrix}$$

$$\hat{\theta}_{ij} = \begin{bmatrix} 2.807 & 1.256 & 1.256 & 1.256 \\ 1.256 & 2.807 & 1.256 & 1.256 \\ 1.256 & 1.256 & 2.807 & 1.256 \\ 1.256 & 1.256 & 1.256 & 2.807 \end{bmatrix}$$

The local odds ratio for the diagonal cells is a common value of 2.807, whereas the off-diagonal is 1.256. The interpretation of the odds ratio of 1.256 is that the odds that a son's status is $j + 1$ instead of j is estimated to be 1.256 times higher when the father's status is $i + 1$ rather than when it is i. For instance, for father's status 1, the odds of son's status at 5 to status at 4 is 0.251 (1.87/7.46), while for father's status 2, the odds of son's status at 5 to status at 4 is 0.316 (20.60/65.23). So the odds ratio of father's status is 2 rather than the status 1 is 1.256 (0.316/0.251).

Goodman (1979a) and Lawal (1992) suggest using estimated odds of row and column adjacent cells to examine the relationship. Two sets of odds are suggested, namely $\Omega_{ij}^{\overline{AB}}$ and $\Omega_{ij}^{\overline{BA}}$. $\Omega_{ij}^{\overline{AB}}$ is defined as the odds that an estimated value falls in column j rather than in column $j + 1$, given that it is in row i, and $\Omega_{ij}^{\overline{BA}}$ is defined as the odds that an estimated value falls in row i rather than in row $i + 1$, given that it is in column j.

$$\Omega_{ij}^{\overline{AB}} = \frac{\hat{m}_{ij}}{\hat{m}_{i,j+1}} \quad \text{and} \quad \Omega_{ij}^{\overline{BA}} = \frac{\hat{m}_{ij}}{\hat{m}_{i+1,j}}$$

From Table 6.12, by examining the output of $\Omega_{ij}^{\overline{AB}}$, we observe that there is a decreasing in odds for all the columns. For instance, the odds of column 1 (1/2) decreases from 0.88 (row 1) to 0.16 (row 5). This implies that the odds of son from status 1 to status 2 (indicated by the column 1/2), the odds decreases if father is at higher status. Similarly, for column two (2/3), the odds of son from status 2 to status 3, the odds also decreases if father is at higher status. This is another way of looking at the odds ratio. In this instance, the odds that a son's status is j instead of $j + 1$ is lower when the father's status is $i + 1$ rather than when it is i. The same observation applies to the rest of the columns (Table 6.13).

The changing of the odds for the same column is according to the estimated odds ratios. For instance, the ratio of 0.88/0.31 = 2.807, the ratio of 0.31/0.25 = 1.256, the ratio of 0.25/0.20 = 1.256, and the ratio of 0.20/0.16 = 1.256.

Table 6.12 $\Omega_{ij}^{\overline{AB}}$ Danish father–son pairs

$\Omega_{ij}^{\overline{AB}}$	1/2	2/3	3/4	4/5
1	0.88	1.12	2.06	3.99
2	0.31	0.89	1.64	3.17
3	0.25	0.32	1.31	2.52
4	0.20	0.25	0.47	2.01
5	0.16	0.20	0.37	0.72

Table 6.13 $\Omega_{ij}^{\overline{BA}}$ Danish father–son pairs

$\Omega_{ij}^{\overline{BA}}$	1	2	3	4	5
1/2	0.51	0.18	0.14	0.11	0.09
2/3	1.31	1.05	0.37	0.30	0.24
3/4	2.52	2.00	1.60	0.57	0.45
4/5	4.13	3.29	2.62	2.08	0.74

The summary of the uniform fixed distance association model is as follows (Lawal 1992).

(a) When a father's status is i and the son's status has the higher status category, the son's status tends to be j rather than $j + 1$, and
(b) when a son's status is j and the father's status has the lower status category, the father's status tends to be $i + 1$ rather than i.

6.4.5 Uniform Variable Distance Association Model

Similar to the last section, the uniform association model and variable distance model can be combined to form a new model. Lawal (1992) designates it as the uniform variable distance model (UV). This model has $(a^2 - 3a + 1)$ degrees of freedom. The multiplicative form of the model in terms of its expected cell frequencies is stated below.

$$m_{ij} = \alpha_i \beta_j \Lambda_{ij} \theta^{ij}$$

$$\Lambda_{ij} = \begin{cases} \prod_{k=i}^{j-1} \delta_k & i < j \\ \prod_{k=j}^{i-1} \delta_k & i > j \end{cases}$$

The symbolic table representation of uniform variable distance association model is shown below. The diagonal cells contain different odds ratios, whereas the off-diagonal cells have a common odds ratio θ_2. In comparison to uniform fixed distance association model discussed in the last section, the difference is on the diagonal odds ratio cells. The former has a constant odds ratio, where the latter has different estimates of odds ratios (Fig. 6.12).

The nonstandard log-linear model for uniform variable distance association model is specified as follows:

$$\ell_{ij} = \mu + \lambda_i^R + \lambda_j^C + \lambda_{ij}^U + \lambda_{ij}^{V1} + \lambda_{ij}^{V2} + \lambda_{ij}^{V3} + \lambda_{ij}^{V4}$$

The factors of the above are defined previously.

Table 6.14 is a well-analyzed social mobility data about paired of British son's and father's occupation status.

The R syntax to generate uniform variable distance association model for the above table is stated below.

```
BritishSonFather <- c (50,  45,  8,   18,   8,
                       28, 174, 84,  154,  55,
                       11,  78, 110, 223,  96,
                       14, 150, 185, 714, 447,
                        3,  42,  72, 320, 411)
V1 <- c(2, 1, 1, 1, 1,
        1, 2, 2, 2, 2,
        1, 2, 2, 2, 2,
        1, 2, 2, 2, 2,
        1, 2, 2, 2, 2)
V1 <- as.factor(V1)
V2 <- c(3, 3, 1, 1, 1,
        3, 3, 1, 1, 1,
        1, 1, 3, 3, 3,
        1, 1, 3, 3, 3,
        1, 1, 3, 3, 3)
V2 <- as.factor(V2)
V3 <- c(4, 4, 4, 1, 1,
        4, 4, 4, 1, 1,
        4, 4, 4, 1, 1,
        1, 1, 1, 4, 4,
        1, 1, 1, 4, 4)
V3 <- as.factor(V3)
```

Fig. 6.12 Uniform variable distance association model

	1	2	3	4
1	θ_{11}	θ_2	θ_2	θ_2
2	θ_2	θ_{22}	θ_2	θ_2
3	θ_2	θ_2	θ_{33}	θ_2
4	θ_2	θ_2	θ_2	θ_{44}

Table 6.14 Occupation
status of British father–son
pairs

Father's status	Son's status				
	1	2	3	4	5
1	50	45	8	18	8
2	28	174	84	154	55
3	11	78	110	223	96
4	14	150	185	714	447
5	3	42	72	320	411

```
V4 <- c(5, 5, 5, 5, 1,
        5, 5, 5, 5, 1,
        5, 5, 5, 5, 1,
        5, 5, 5, 5, 1,
        1, 1, 1, 1, 5)
V4 <- as.factor(V4)
UniformVariable <- glm
(BritishSonFather ~ r + c + V1 + V2 + V3 + V4 + U, family=poisson)
summary(UniformVariable)
model.summary(UniformVariable)
```

The fitted values and odds ratios are printed below. The off-diagonal cells have a constant odds ratio of 1.247, whereas the odds ratios are with four different odds ratios.

$$
\hat{m}_{ij} = \begin{bmatrix}
50.00 & 38.51 & 14.00 & 20.19 & 6.30 \\
27.87 & 180.62 & 81.88 & 147.28 & 57.34 \\
10.82 & 87.40 & 96.90 & 217.35 & 105.53 \\
14.06 & 141.71 & 195.93 & 721.47 & 436.82 \\
3.24 & 40.76 & 70.28 & 322.71 & 411.00
\end{bmatrix}
$$

$$
\hat{\theta}_{ij} = \begin{bmatrix}
8.414 & 1.247 & 1.247 & 1.247 \\
1.247 & 2.446 & 1.247 & 1.247 \\
1.247 & 1.247 & 1.642 & 1.247 \\
1.247 & 1.247 & 1.247 & 2.104
\end{bmatrix}
$$

Table 6.15 $\Omega_{ij}^{\overline{AB}}$ British
father–son pairs

$\Omega_{ij}^{\overline{AB}}$	1/2	2/3	3/4	4/5
1	1.30	2.75	0.69	3.20
2	0.15	2.21	0.56	2.57
3	0.12	0.90	0.45	2.06
4	0.10	0.72	0.27	1.65
5	0.08	0.58	0.22	0.79

The estimated values of $\Omega_{ij}^{\overline{AB}}$ and $\Omega_{ij}^{\overline{BA}}$ are printed below (Tables 6.15 and 6.16).

Exercises

6.1 Write a R function to generate the L factor for an $a \times a$ table using the following formula.

$$L = \begin{cases} 1 & i \neq j \\ 2 & i = j \end{cases}$$

6.2 Write a R function to generate the Q factor for an $a \times a$ table using the following formula.

$$Q = \begin{cases} 1 & i \neq j \\ i+1 & i = j \end{cases}$$

6.3 Fit the non-independence models discussed in this chapter for the Table 6.17. This a 6×6 table taken from Hutchinson (1958), a common table analyzed by many authors. It is about Brazilian intergenerational mobility.

6.4 The following is a 5×5 British social mobility table of father's and son's status (Bishop et al. 1975). Fit the fixed distance, variable distance model, uniform association model, uniform fixed distance model, and uniform variable distance model for the Table 6.18.

6.5 The following is a 5×5 Danish social mobility table of father's and son's status. Fit the fixed distance, variable distance model, uniform association model, uniform fixed distance model, and uniform variable distance model for the Table 6.19.

Table 6.16 $\Omega_{ij}^{\overline{BA}}$ British father–son pairs

$\Omega_{ij}^{\overline{BA}}$	1	2	3	4	5
1/2	1.79	0.21	0.17	0.14	0.11
2/3	2.58	2.07	0.84	0.68	0.54
3/4	0.77	0.62	0.49	0.30	0.24
4/5	4.34	3.48	2.79	2.24	1.06

Table 6.17 Father's and son's occupation status

Father's occupation	Son's occupation					
	1	2	3	4	5	6
1	33	12	10	3	0	0
2	14	25	16	3	2	0
3	13	16	68	21	16	1
4	6	30	39	74	61	7
5	5	16	26	45	132	24
6	1	9	29	41	142	116

Table 6.18 Father's and son's occupation status

Father's status	Son's status				
	1	2	3	4	5
1	50	45	8	18	8
2	28	174	84	154	55
3	11	78	110	223	96
4	14	150	185	714	447
5	3	42	72	320	411

Table 6.19 Father's and son's occupation status

Father's status	Son's status				
	1	2	3	4	5
1	18	17	16	4	2
2	24	105	109	59	21
3	23	84	289	217	95
4	8	49	175	384	198
5	6	8	69	201	246

References

Bishop, Y. M., Fienberg, S. E., & Holland, P. W. (1975). *Discrete multivariate analysis: Theory and applications*. Berlin: Springer.

Goodman, L. A. (1972). Some multiplicative models for the analysis of cross classified data. In L. Le Cam, J. Neyman & E. L. Scott (Eds.), *Proceedings of the sixth Berkeley symposium on mathematical statistics and probability*. Berkeley: University of California Press.

Goodman, L. A. (1979a). Simple models for the analysis of association in cross-classifications having ordered categories. *Journal of American Statistical Association, 74*, 537–552.

Goodman, L. A. (1979b). Multiplicative models for the analysis of occupational mobility tables and other kinds of cross-classification tables. *The American Journal of Sociology, 84*(4), 804–819.

Goodman, L. A. (1996). A single general method for the analysis of cross-classified data: Reconciliation and synthesis of some methods of Pearson, Yule, and Fisher, and some methods of correspondence analysis and association analysis. *Journal of the American Statistical Association, 91*(433), 408–428.

Hutchinson, B. (1958). Structural and exchange mobility in the assimilation of immigrants to Brazil. *Population Studies, 12*, 111–120.

Haberman, S. J. (1974). *The analysis of frequency data*. Chicago: University of Chicago Press.

Haberman, S. J. (1979). *Analysis of qualitative data*, Vol. 2. New Development, Academic Press.

Lawal, H. B. (1992). Parsimonious uniform-distance association models for the occupational mobility data. *Journal of the Japan Statistical Society, 22*(2), 183–192.

Lawal, H. B. (2001). Modeling symmetry models in square contingency tables with ordered categories. *Journal of Statistical Computation and Simulation, 71*, 59–83.

Lawal, H. B. (2003). *Categorical data analysis with SAS and SPSS applications*. Lawrence Erlbaum Associates, Publishers.

Lawal, H. B. (2004). Using a GLM to decompose the symmetry model in square contingency tables with ordered categories. *Journal of Applied Statistics, 31*(3), 279–303.

Scheuren, J. K., & Loch O. H. (1975). A data analysis approach to fitting square tables. *Communications in Statistics, 4*, 595–615.

Upton, G. J. G., & Sarvik, B. (1981). A loyalty-distance model for voting change. *Journal of Royal Statistical Society, A, 144*(2), 247–259.

Upton, G. J. G. (1985). A survey of log linear models for ordinal variables in an I × J contingency table. *Guru Nanak Journal of Sociology, 6*, 1–18.

Chapter 7
Asymmetry + Non-Independence Models

7.1 Non-Symmetry + Independence Model

The base model for the category of asymmetry + non-independence models is the null non-symmetry + independence model. Hope (1982) refers it as the Halfway model. The multiplicative form for the model is stated below (Goodman 1985; Lawal 2004).

$$\pi_{ij} = \alpha_i \alpha_j \text{ for } (i \neq j)$$

The symbolic table for the null non-symmetry + independence model is printed below. The odds ratios are all one for this model (Fig. 7.1).

The nonstandard log-linear formulation of null non-symmetry + independence model is as follows:

$$\ell_{ij} = \mu + \sum_{i=1}^{I-1} \beta_i H_i$$

where H_i is the series of regression variables defined below (Hope 1982; Lawal 2004). For a 6×6 table, H_i are generated from H_1 up to H_5, and for a 7×7 table, H_i are generated from H_1 up to H_6. In general, for an $a \times a$ table, H_i are generated from H_1 up to H_{a-1}.

$$f_i = \begin{cases} 1 & \text{if cell in row i} \\ 0 & \text{otherwise} \end{cases}, s_i = \begin{cases} 1 & \text{if cell in row i} \\ 0 & \text{otherwise} \end{cases} \text{ and}$$
$$H_i = f_i + s_i \quad i = 1, 2, \ldots, I-1$$

The following illustrates, for a 6×6 table, the entries of $f_1, s_1, H_1 f_2, s_2,$ and H_2.

© Springer Nature Singapore Pte Ltd. 2017
T.K. Tan, *Doubly Classified Model with R*,
https://doi.org/10.1007/978-981-10-6995-6_7

θ	1	2	3	4
1	1	1	1	1
2	1	1	1	1
3	1	1	1	1
4	1	1	1	1

Fig. 7.1 Non-symmetry + independence model

$$f_1 = \begin{pmatrix} 1 & 1 & 1 & 1 & 1 & 1 \\ 0 & 0 & 0 & 0 & 0 & 0 \\ 0 & 0 & 0 & 0 & 0 & 0 \\ 0 & 0 & 0 & 0 & 0 & 0 \\ 0 & 0 & 0 & 0 & 0 & 0 \\ 0 & 0 & 0 & 0 & 0 & 0 \end{pmatrix}, s_1 = \begin{pmatrix} 1 & 0 & 0 & 0 & 0 & 0 \\ 1 & 0 & 0 & 0 & 0 & 0 \\ 1 & 0 & 0 & 0 & 0 & 0 \\ 1 & 0 & 0 & 0 & 0 & 0 \\ 1 & 0 & 0 & 0 & 0 & 0 \\ 1 & 0 & 0 & 0 & 0 & 0 \end{pmatrix},$$

$$H_1 = \begin{pmatrix} 2 & 1 & 1 & 1 & 1 & 1 \\ 1 & 0 & 0 & 0 & 0 & 0 \\ 1 & 0 & 0 & 0 & 0 & 0 \\ 1 & 0 & 0 & 0 & 0 & 0 \\ 1 & 0 & 0 & 0 & 0 & 0 \\ 1 & 0 & 0 & 0 & 0 & 0 \end{pmatrix}$$

$$f_2 = \begin{pmatrix} 0 & 0 & 0 & 0 & 0 & 0 \\ 1 & 1 & 1 & 1 & 1 & 1 \\ 0 & 0 & 0 & 0 & 0 & 0 \\ 0 & 0 & 0 & 0 & 0 & 0 \\ 0 & 0 & 0 & 0 & 0 & 0 \\ 0 & 0 & 0 & 0 & 0 & 0 \end{pmatrix}, s_2 = \begin{pmatrix} 0 & 1 & 0 & 0 & 0 & 0 \\ 0 & 1 & 0 & 0 & 0 & 0 \\ 0 & 1 & 0 & 0 & 0 & 0 \\ 0 & 1 & 0 & 0 & 0 & 0 \\ 0 & 1 & 0 & 0 & 0 & 0 \\ 0 & 1 & 0 & 0 & 0 & 0 \end{pmatrix},$$

$$H_2 = \begin{pmatrix} 0 & 1 & 0 & 0 & 0 & 0 \\ 1 & 2 & 1 & 1 & 1 & 1 \\ 0 & 1 & 0 & 0 & 0 & 0 \\ 0 & 1 & 0 & 0 & 0 & 0 \\ 0 & 1 & 0 & 0 & 0 & 0 \\ 0 & 1 & 0 & 0 & 0 & 0 \end{pmatrix}$$

The R syntax for generating H_i for analyzing a 6×6 table is given below.

```
f1 <- c(1, 1, 1, 1, 1, 1,
        0, 0, 0, 0, 0, 0,
        0, 0, 0, 0, 0, 0,
        0, 0, 0, 0, 0, 0,
```

```
        0, 0, 0, 0, 0, 0,
        0, 0, 0, 0, 0, 0)
s1 <- c(1, 0, 0, 0, 0, 0,
        1, 0, 0, 0, 0, 0,
        1, 0, 0, 0, 0, 0,
        1, 0, 0, 0, 0, 0,
        1, 0, 0, 0, 0, 0,
        1, 0, 0, 0, 0, 0)
f2 <- c(0, 0, 0, 0, 0, 0,
        1, 1, 1, 1, 1, 1,
        0, 0, 0, 0, 0, 0,
        0, 0, 0, 0, 0, 0,
        0, 0, 0, 0, 0, 0,
        0, 0, 0, 0, 0, 0)
s2 <- c(0, 1, 0, 0, 0, 0,
        0, 1, 0, 0, 0, 0,
        0, 1, 0, 0, 0, 0,
        0, 1, 0, 0, 0, 0,
        0, 1, 0, 0, 0, 0,
        0, 1, 0, 0, 0, 0)
f3 <- c(0, 0, 0, 0, 0, 0,
        0, 0, 0, 0, 0, 0,
        1, 1, 1, 1, 1, 1,
        0, 0, 0, 0, 0, 0,
        0, 0, 0, 0, 0, 0,
        0, 0, 0, 0, 0, 0)
s3 <- c(0, 0, 1, 0, 0, 0,
        0, 0, 1, 0, 0, 0,
        0, 0, 1, 0, 0, 0,
        0, 0, 1, 0, 0, 0,
        0, 0, 1, 0, 0, 0,
        0, 0, 1, 0, 0, 0)
f4 <- c(0, 0, 0, 0, 0, 0,
        0, 0, 0, 0, 0, 0,
        0, 0, 0, 0, 0, 0,
        1, 1, 1, 1, 1, 1,
        0, 0, 0, 0, 0, 0,
        0, 0, 0, 0, 0, 0)
s4 <- c(0, 0, 0, 1, 0, 0,
        0, 0, 0, 1, 0, 0,
        0, 0, 0, 1, 0, 0,
        0, 0, 0, 1, 0, 0,
        0, 0, 0, 1, 0, 0,
        0, 0, 0, 1, 0, 0)
f5 <- c(0, 0, 0, 0, 0, 0,
        0, 0, 0, 0, 0, 0,
        0, 0, 0, 0, 0, 0,
```

Table 7.1 Intergenerational mobility table	Father's occupation	Son's occupation					
		1	2	3	4	5	6
	1	29	15	11	7	4	1
	2	15	16	21	12	10	3
	3	11	21	60	36	25	13
	4	7	12	35	70	54	26
	5	4	9	25	53	139	72
	6	1	4	15	26	75	128

```
        0, 0, 0, 0, 0, 0,
        1, 1, 1, 1, 1, 1,
        0, 0, 0, 0, 0, 0)
s5 <- c(0, 0, 0, 0, 1, 0,
        0, 0, 0, 0, 1, 0,
        0, 0, 0, 0, 1, 0,
        0, 0, 0, 0, 1, 0,
        0, 0, 0, 0, 1, 0,
        0, 0, 0, 0, 1, 0)
h1 <- f1 + s1
h2 <- f2 + s2
h3 <- f3 + s3
h4 <- f4 + s4
h5 <- f5 + s5
```

Table 7.1 below prints the intergenerational mobility table between son's and father's occupation. This example will be used for illustration for the entire chapter.

The R syntax for generating the null non-symmetry + independence model for Table 7.1 is stated below.

```
# Null Model: Non-Symmetry + Independence
Mobility <- c(29, 17, 11,  7,   4,   1,
              15, 16, 21, 12,  10,   3,
              12, 23, 60, 36,  27,  13,
               7, 17, 35, 70,  54,  26,
               4,  9, 25, 53, 139,  72,
               1,  4, 15, 26,  75, 128)
ANNull <- glm(Mobility ~ h1+h2+h3+h4+h5, family=poisson)
summary(ANNull)
model.summary(ANNull)
```

Table 7.2 below lists the estimated values.

```
[1] "Local Odds Ratios Table"
     [,1] [,2] [,3] [,4] [,5]
[1,]    1    1    1    1    1
[2,]    1    1    1    1    1
[3,]    1    1    1    1    1
[4,]    1    1    1    1    1
[5,]    1    1    1    1    1
```

The fitted odds ratios of non-symmetry + independence model are all equal to one.

Table 7.2 Fitted values—Non-symmetry + independence model

Father's occupation	Son's occupation					
	1	2	3	4	5	6
1	4.36	5.18	10.75	13.13	19.43	15.65
2	5.18	6.17	12.79	15.63	23.12	18.62
3	10.75	12.79	26.52	32.40	47.94	38.60
4	13.13	15.63	32.40	39.59	58.58	47.17
5	19.43	23.12	47.94	58.58	86.66	69.78
6	15.65	18.62	38.60	47.17	69.78	56.19

7.2   Non-Symmetry + Independence Triangle Model

Last section covers a composite model by combining non-symmetry and independence model. This section introduces non-symmetry + independence triangle model. As the name suggests, it is a combination of non-symmetric and the triangle parameters model. The multiplicative form of non-symmetry + independence triangle model is stated below (Goodman 1985; Lawal 2004).

$$\pi_{ij} = \begin{cases} \alpha_i \alpha_j \tau_1 & i > j \\ \alpha_i \alpha_j \tau_2 & i < j \end{cases}$$

The symbolic table below shows the main characteristics of non-symmetry + independence triangle model. The main diagonal of the odds ratios table has a common odds ratio θ_1. The upper and the lower diagonals with one position away from the diagonal each has a common odds ratio of θ_2 and θ_3, respectively. The rest of the odds ratios are of a value of one (Fig. 7.2).

The nonstandard form of non-symmetry + independence triangle model is as follows:

$$\ell_{ij} = \mu + \sum_{i=1}^{I-1} \beta_i H_i + \lambda_{ij}^T$$

	1	2	3	4
1	θ_1	θ_2	1	1
2	θ_3	θ_1	θ_2	1
3	1	θ_3	θ_1	θ_2
4	1	1	θ_3	θ_1

Fig. 7.2 Non-symmetry + independence triangle model

$$\text{where } T = \begin{pmatrix} 3 & 1 & 1 & 1 & 1 & 1 \\ 2 & 3 & 1 & 1 & 1 & 1 \\ 2 & 2 & 3 & 1 & 1 & 1 \\ 2 & 2 & 2 & 3 & 1 & 1 \\ 2 & 2 & 2 & 2 & 3 & 1 \\ 2 & 2 & 2 & 2 & 2 & 3 \end{pmatrix}$$

The R syntax for fitting the model is as follows:

```
# Non-Symmetry+Independence Traingles Model
ANT <- glm(Mobility ~ h1+h2+h3+h4+h5+T, family=poisson)
summary(ANT)
model.summary(ANT)
```

The fitted values and odds ratios are presented in Tables 7.3 and 7.4, respectively.

Table 7.3 Fitted values—Non-symmetry + independence triangle model

Father's occupation	Son's occupation					
	1	2	3	4	5	6
1	13.22	5.24	9.68	11.36	15.36	13.02
2	5.36	17.96	11.28	13.24	17.90	15.18
3	9.90	11.53	61.29	24.46	33.06	28.04
4	11.61	13.54	25.00	84.39	38.80	32.90
5	15.70	18.30	33.80	39.66	154.23	44.48
6	13.31	15.52	28.66	33.63	45.47	110.90

Table 7.4 Fitted odds ratios —Non-symmetry + independence triangle model

Θ	1	2	3	4	5
1	8.46	0.34	1.00	1.00	1.00
2	0.35	8.46	0.34	1.00	1.00
3	1.00	0.35	8.46	0.34	1.00
4	1.00	1.00	0.35	8.46	0.34
5	1.00	1.00	1.00	0.35	8.46

7.3 Non-Symmetry + Independence Diagonals Model

The non-symmetry + independence diagonal model is the combination of non-symmetric and the diagonal parameters model. The multiplicative form of non-symmetry + independence diagonal model is stated below (Goodman 1985; Lawal 2004). The symbolic table below shows that the odds ratios are in parallel diagonally, and each diagonal has a common odds ratio. For a 4×4 odds ratio table, there are seven odds ratios (Fig. 7.3).

$$\pi_{ij} = \alpha_i \alpha_j \delta_k \text{ for } i \neq j, k = i - j$$

The nonstandard log-linear form of non-symmetry + independence diagonal model is stated as follows:

$$\ell_{ij} = \mu + \sum_{i=1}^{I-1} \beta_i H_i + \lambda_{ij}^D$$

$$\text{where } D = \begin{pmatrix} 11 & 1 & 2 & 3 & 4 & 5 \\ 6 & 11 & 1 & 2 & 3 & 4 \\ 7 & 6 & 11 & 1 & 2 & 3 \\ 8 & 7 & 6 & 11 & 1 & 2 \\ 9 & 8 & 7 & 6 & 11 & 1 \\ 10 & 9 & 8 & 7 & 6 & 11 \end{pmatrix}$$

The R syntax to generate non-symmetry + independence triangle model is as follows:

```
# Non-Symmetry+Independence Diagonals Model
AND <- glm(Mobility ~ h1+h2+h3+h4+h5+D, family=poisson)
summary(AND)
model.summary(AND)
```

The fitted values and odds ratio are shown in Tables 7.5 and 7.6 below.

	1	2	3	4
1	θ_1	θ_2	θ_3	θ_4
2	θ_5	θ_1	θ_2	θ_3
3	θ_6	θ_5	θ_1	θ_2
4	θ_7	θ_6	θ_5	θ_1

Fig. 7.3 Non-symmetry + independence diagonal model

Table 7.5 Fitted values—Non-symmetry + independence diagonals model

Father's occupation	Son's occupation					
	1	2	3	4	5	6
1	29.27	14.46	11.99	7.23	3.81	1.00
2	14.53	22.34	20.56	11.30	8.89	3.19
3	12.62	20.66	59.20	36.10	25.90	13.88
4	7.47	11.90	36.28	68.89	54.83	26.81
5	4.35	9.19	27.27	55.10	136.58	74.06
6	1.00	3.65	14.35	28.22	74.43	125.71

Table 7.6 Estimated odds ratios —Non-symmetry + independence diagonals model

Θ	1	2	3	4	5
1	3.11	1.11	0.91	1.49	1.37
2	1.06	3.11	1.11	0.91	1.49
3	0.97	1.06	3.11	1.11	0.91
4	1.33	0.97	1.06	3.11	1.11
5	1.73	1.33	0.97	1.06	3.11

7.4 Non-Symmetry + Independence Diagonals Absolute Model

The non-symmetry + independence diagonals absolute model is a combination of non-symmetric and the diagonals absolute model. The multiplicative form of this model is stated below (Goodman 1985; Lawal 2004).

$$\pi_{ij} = \alpha_i \alpha_j \delta'_k \text{ for } i \neq j, k' = |i - j|$$

The symbolic table below shows that the main diagonal has a common odds ratio. The off-diagonal cells also have a common odds ratio for those cells which have the same distance away from the diagonal. For a 4×4 odds ratio table, there are four odds ratios (Fig. 7.4).

	1	2	3	4
1	θ_1	θ_2	θ_3	θ_4
2	θ_2	θ_1	θ_2	θ_3
3	θ_3	θ_2	θ_1	θ_2
4	θ_4	θ_3	θ_2	θ_1

Fig. 7.4 Non-symmetry + independence diagonal absolute model

The nonstandard log-linear has the following specification.

$$\ell_{ij} = \mu + \sum_{i=1}^{I-1} \beta_i H_i + \lambda_{ij}^{DA}$$

$$\text{where } DA = \begin{pmatrix} 6 & 1 & 2 & 3 & 4 & 5 \\ 1 & 6 & 1 & 2 & 3 & 4 \\ 2 & 1 & 6 & 1 & 2 & 3 \\ 3 & 2 & 1 & 6 & 1 & 2 \\ 4 & 3 & 2 & 1 & 6 & 1 \\ 5 & 4 & 3 & 2 & 1 & 6 \end{pmatrix}$$

The R syntax is stated below.

```
# Non-Symmetry+Independence Diagonals Absolute Model
ANDA <- glm(Mobility ~ h1+h2+h3+h4+h5+DA, family=poisson)
summary(ANDA)
model.summary(ANDA)
```

Tables 7.7 and 7.8 print the fitted values and odds ratios, respectively.

Table 7.7 Fitted values—Non-symmetry + independence diagonals absolute model

Father's occupation	Son's occupation					
	1	2	3	4	5	6
1	29.27	14.49	12.31	7.35	4.08	1.00
2	14.49	22.34	20.61	11.60	9.04	3.42
3	12.31	20.61	59.20	36.19	26.58	14.11
4	7.35	11.60	36.19	68.89	54.97	27.51
5	4.08	9.04	26.58	54.97	136.58	74.25
6	1.00	3.42	14.11	27.51	74.25	125.71

Table 7.8 Fitted odds ratios —Non-symmetry + independence diagonals absolute model

Θ	1	2	3	4	5
1	3.11	1.09	0.94	1.40	1.54
2	1.09	3.11	1.09	0.94	1.40
3	0.94	1.09	3.11	1.09	0.94
4	1.40	0.94	1.09	3.11	1.09
5	1.54	1.40	0.94	1.09	3.11

7.5   Non-Symmetry + Independence Diagonals Absolute Triangle Model

The last model of this set of models is the non-symmetry + independence diagonals absolute triangle model. It is the combination of non-symmetric and the diagonals absolute triangle model. The multiplicative form of this model is stated below (Goodman 1985; Lawal 2004).

$$\pi_{ij} = \begin{cases} \alpha_i \alpha_j \delta'_k \tau_1 & i > j \\ \alpha_i \alpha_j \delta'_k \tau_2 & i < j \end{cases}$$

The symbolic table shows that all diagonals have a common odds ratio. The odds ratio for those cells one position away from the diagonal each has a common odds ratio, separately for upper and lower diagonal. For instance, the odd ratio of cell (1, 2), cell (2, 3), cell (3, 4) has a common odds ratio θ_2 which is different from the odd ratio of the lower diagonal of cell (2, 1), cell (3, 2), and cell (4, 3) with another common odds ratio of θ_3. Those cells two positions away from the diagonal have a common odds ratio, both for upper and lower diagonal. For instance, cells that are two position away have a common odds ratio of θ_4 and cells that are three positions away have another common odds ratio of θ_5. So, altogether for a 4 × 4 table, there are five odds ratios (Fig. 7.5).

The nonstandard log-linear formulation of null non-symmetry + independence diagonal absolute triangle model is as follows:

$$\ell_{ij} = \mu + \sum_{i=1}^{I-1} \beta_i H_i + \lambda_{ij}^{DA} + \lambda_{ij}^{T}$$

The R syntax for the model is stated below.

```
# Non-Symmetry+Independencen Diagonals Absolute Traingle Model
ANDAT <- glm(Mobility ~ h1+h2+h3+h4+h5+T+DA, family=poisson)
summary(ANDAT)
model.summary(ANDAT)
```

	1	2	3	4
1	θ_1	θ_2	θ_4	θ_5
2	θ_3	θ_1	θ_2	θ_4
3	θ_4	θ_3	θ_1	θ_2
4	θ_5	θ_4	θ_3	θ_1

Fig. 7.5 Non-symmetry + independence diagonal absolute triangle model

Table 7.9 Fitted values—Non-symmetry + independence diagonals absolute triangle model

Father's occupation	Son's occupation					
	1	2	3	4	5	6
1	29.27	14.33	12.17	7.27	4.04	0.99
2	14.65	22.34	20.38	11.47	8.94	3.38
3	12.44	20.84	59.20	35.79	26.29	13.96
4	7.43	11.73	36.59	68.89	54.36	27.21
5	4.13	9.14	26.88	55.57	136.58	73.43
6	1.01	3.46	14.27	27.81	75.07	125.71

Table 7.10 Fitted odds ratios—Non-symmetry + independence diagonals absolute triangle model

Θ	1	2	3	4	5
1	3.11	1.07	0.94	1.40	1.54
2	1.10	3.11	1.07	0.94	1.40
3	0.94	1.10	3.11	1.07	0.94
4	1.40	0.94	1.10	3.11	1.07
5	1.54	1.40	0.94	1.10	3.11

The following two tables produce the fitted values and odds ratios for non-symmetry + independence diagonal absolute triangle model (Tables 7.9 and 7.10).

The fit statistics for all the asymmetric + non-independence models so far discussed are summarized in Table 7.11 below. The first three models (null, independence, and triangle) do not fit well while the last three models have good fit. The results of AIC and BIC show that diagonals absolute model has the best fit.

Table 7.11 Non-symmetry + non-independence model

Model	Df	P-value	G^2	AIC	BIC
Null	30	0.0000	495.96	435.96	286.50
Independence	25	0.0000	495.19	445.19	320.64
Triangle	28	0.0000	242.78	186.78	47.29
Diagonals	20	0.9999	5.90	−34.10	−133.74
Diagonals absolute	25	1.0000	6.09	−43.91	−168.46
Diagonals absolute triangle	24	0.9999	6.01	−41.99	−161.56

7.6 Non-Symmetry + Independence Models—Without Diagonal Cells

The models discussed in the previous six sections can be modeled without taking into account the diagonal cells. The following R syntax shows the fitting of the models without the diagonal cells. The diagonal cells are removed from the vector MobiityNoD resulting in a 5×5 table. The specifications of the rest are as follows.

```
MobilityNoD <- c(17, 11,  7,  4,  1,
                 15, 21, 12, 10,  3,
                 12, 23, 36, 27, 13,
                  7, 17, 35, 54, 26,
                  4,  9, 25, 53, 72,
                  1,  4, 15, 26, 75)
f1 <- c(1, 1, 1, 1, 1,
        0, 0, 0, 0, 0,
        0, 0, 0, 0, 0,
        0, 0, 0, 0, 0,
        0, 0, 0, 0, 0,
        0, 0, 0, 0, 0)
s1 <- c(0, 0, 0, 0, 0,
        1, 0, 0, 0, 0,
        1, 0, 0, 0, 0,
        1, 0, 0, 0, 0,
        1, 0, 0, 0, 0,
        1, 0, 0, 0, 0)
f2 <- c(0, 0, 0, 0, 0,
        1, 1, 1, 1, 1,
        0, 0, 0, 0, 0,
        0, 0, 0, 0, 0,
        0, 0, 0, 0, 0,
        0, 0, 0, 0, 0)
s2 <- c(1, 0, 0, 0, 0,
        0, 0, 0, 0, 0,
        0, 1, 0, 0, 0,
        0, 1, 0, 0, 0,
        0, 1, 0, 0, 0,
        0, 1, 0, 0, 0)
f3 <- c(0, 0, 0, 0, 0,
        0, 0, 0, 0, 0,
        1, 1, 1, 1, 1,
        0, 0, 0, 0, 0,
        0, 0, 0, 0, 0,
        0, 0, 0, 0, 0)
```

```
s3 <- c(0, 1, 0, 0, 0,
        0, 1, 0, 0, 0,
        0, 0, 0, 0, 0,
        0, 0, 1, 0, 0,
        0, 0, 1, 0, 0,
        0, 0, 1, 0, 0)
f4 <- c(0, 0, 0, 0, 0,
        0, 0, 0, 0, 0,
        0, 0, 0, 0, 0,
        1, 1, 1, 1, 1,
        0, 0, 0, 0, 0,
        0, 0, 0, 0, 0)
s4 <- c(0, 0, 1, 0, 0,
        0, 0, 1, 0, 0,
        0, 0, 1, 0, 0,
        0, 0, 0, 0, 0,
        0, 0, 0, 1, 0,
        0, 0, 0, 1, 0)
f5 <- c(0, 0, 0, 0, 0,
        0, 0, 0, 0, 0,
        0, 0, 0, 0, 0,
        0, 0, 0, 0, 0,
        1, 1, 1, 1, 1,
        0, 0, 0, 0, 0)
s5 <- c(0, 0, 0, 1, 0,
        0, 0, 0, 1, 0,
        0, 0, 0, 1, 0,
        0, 0, 0, 1, 0,
        0, 0, 0, 0, 0,
        0, 0, 0, 0, 1)
h1 <- f1 + s1
h2 <- f2 + s2
h3 <- f3 + s3
h4 <- f4 + s4
h5 <- f5 + s5
T <- c(1, 1, 1, 1, 1,
       2, 1, 1, 1, 1,
       2, 2, 1, 1, 1,
       2, 2, 2, 1, 1,
       2, 2, 2, 2, 1,
       2, 2, 2, 2, 2)
T <- as.factor(T)
D <- c( 1,  2,  3,  4,  5,
        6,  1,  2,  3,  4,
        7,  6,  1,  2,  3,
```

```
        8,  7,  6,  1,  2,
         9,  8,  7,  6,  1,
        10,  9,  8,  7,  6)
D <- as.factor(D)
DA <- c(1, 2, 3, 4, 5,
          1, 1, 2, 3, 4,
          2, 1, 1, 2, 3,
          3, 2, 1, 1, 2,
          4, 3, 2, 1, 1,
          5, 4, 3, 2, 1)
DA <- as.factor(DA)
```

```
# Null Model: Non-Symmetry+Independence
NoDANNull <- glm(MobilityNoD ~ h1+h2+h3+h4+h5, family=poisson)
summary(NoDANNull)
model.summary(NoDANNull)
matrix(fitted(NoDANNull),nrow=6,ncol=6,byrow=T)
# Non-Symmetry+Independencen Traingles Model
NoDANT <- glm(MobilityNoD ~ h1+h2+h3+h4+h5+T, family=poisson)
summary(NoDANT)
model.summary(NoDANT)
matrix(fitted(NoDANT),nrow=6,ncol=5,byrow=T)
# Non-Symmetry+Independencen Diagonals Model
NoDAND <- glm(MobilityNoD ~ h1+h2+h3+h4+h5+D, family=poisson)
summary(NoDAND)
model.summary(NoDAND)
matrix(fitted(NoDAND),nrow=6,ncol=5,byrow=T)
# Non-Symmetry+Independencen Diagonals Absolute Model
NoDANDA <- glm(MobilityNoD ~ h1+h2+h3+h4+h5+DA, family=poisson)
summary(NoDANDA)
model.summary(NoDANDA)
matrix(fitted(NoDANDA),nrow=6,ncol=5,byrow=T)
# Non-Symmetry+Independencen Diagonals Absolute Traingle Model
NoDANDAT <- glm(MobilityNoD ~ h1+h2+h3+h4+h5+T+DA, family=poisson)
summary(NoDANDAT)
model.summary(NoDANDAT)
matrix(fitted(NoDANDAT),nrow=6,ncol=5,byrow=T)
```

The fit of the models for non-symmetry and non-independence models without the consideration of diagonal cells is summarized in Table 7.12 below. According to AIC and BIC, diagonals absolute model fits best. This result is consistent with

Table 7.12 Non-symmetry + non-independence models (without cells at diagonal)

Model	Df	P-value	G^2	AIC	BIC
Null	24	0.0000	195.95	147.95	41.07
Triangle	23	0.0000	195.88	149.88	47.44
Diagonals	15	1.0000	2.07	−27.93	−94.74
Diagonals absolute	20	1.0000	2.25	−37.75	−126.82
Diagonals absolute triangle	19	1.0000	2.18	−35.82	−120.44

Table 7.11 where the diagonal cells are considered for modeling. In general, for doubly classified modeling, if the diagonal cells are not relevant for analysis, this approach of removing diagonal cells can be adopted.

Exercises

7.1 The following is a 6 × 6 table of Sao Paulo, Brazil, intergenerational mobility table based on son and father occupation (Hutchinson 1958; Lawal 2004). This is an extensively analyzed table by many authors. Fit null asymmetry non-independence model, asymmetry non-independence triangle model, asymmetry non-independence diagonals model, asymmetry non-independence diagonals absolute model, and asymmetry non-independence diagonals absolute triangle model. Comment on the results (Table 7.13).

7.2 Fit the same set of asymmetry non-independence model as stated in 7.1 but removing the diagonal cells. Comment on the results.

Table 7.13 Father's and son's occupation

Father's occupation	Son's occupation					
	1	2	3	4	5	6
1	33	12	10	3	0	0
2	14	25	16	3	2	0
3	13	16	68	21	16	1
4	6	30	39	74	61	7
5	5	16	26	45	132	24
6	1	9	29	41	142	116

References

Goodman, L. A. (1985). The analysis of cross-classified data having ordered and/or unordered categories: Association models, correlation models, and asymmetry models for contingency tables with or without missing entries. *The Annals of Statistics 13,* 10–69.

Hope, K. (1982). Vertical and non vertical class mobility in three countries. *American Sociological Review, 47,* 99–113.

Hutchinson, B. (1958). Structural and exchange mobility in the assimilation of immigrants to Brazil. *Population Studies, 12,* 111–120.

Lawal, H. B. (2003). *Categorical Data Analysis with SAS and SPSS Applications.* Lawrence Erlbaum Associates, Publishers.

Lawal, H. B. (2004). Review of non-independence, asymmetry, skew-symmetry and point-symmetry models in the analysis of social mobility data. *Quality and Quantity, 38,* 259–289.

Chapter 8
Modeling Strategy

8.1 Fit Statistics

Using fit statistics to select model is briefly mentioned in Sects. 7.5 and 7.6. This section formally introduces them. The goodness of fit (GOF) of a statistical model describes and gives the statistics to show how well the doubly classified table fits the model. GOF indices summarize the discrepancy between the observed values and the values expected under a statistical model. GOF statistics are indices with known sampling distributions to test for the degree of deviation of the data from the model. Likelihood ratio or deviance and information criterion indices are two main GOF statistics used for evaluation the fit of the data to the model.

Likelihood Ratio/Deviance Statistic
Likelihood ratio or deviance is a common GOF statistic for testing categorical data. Let n_{ij} denotes the observed frequency in the ith row and jth column of a doubly classified table where $i = 1, \ldots, a$; $j = 1, \ldots.a$. The likelihood ratio G^2 statistic for testing goodness of fit of a model is as follows:

$$G^2 = 2 \sum_{i=1}^{r} \sum_{j=1}^{r} n_{ij} \ln \left(\frac{n_{ij}}{\hat{m}_{ij}} \right),$$

where $\hat{m}_{ij}$ is the MLE of expected frequency m_{ij} under the model.

Information Criterion Indices
Likelihood ratio is carried out for models are nested. Information criterion indices are statistics for assessing fit especially useful for comparison of models that are not nested. With L denoting the loglikelihood, two popular GOF indices are Akaike's information criterion (AIC) and Schwarz Bayesian information criterion (BIC). Their formula are stated below.

© Springer Nature Singapore Pte Ltd. 2017
T.K. Tan, *Doubly Classified Model with R*,
https://doi.org/10.1007/978-981-10-6995-6_8

$$AIC = -2L + 2q$$

$$BIC = -2L + q \ln(N)$$

The AIC and BIC are not used to test the model in the sense of hypothesis testing, but for model selection. The model with the lowest value is selected. Notice that both the AIC and BIC combine absolute fit with model parsimony. That is, they penalize by adding parameters to the model, but they do so differently. Of the two statistics, the BIC penalizes by adding parameters to the model more strongly than the AIC.

8.2 Graphical Approach

Doubly classified table can be represented in graphical forms. Instead of examining the numerical figures of a table, a graphical representation of a table will help to identify whether certain doubly classified models fit well. The following subsections introduce two main types of graphical output for doubly classified table, namely heat map square table and three-dimensional heat map bar plot. For examining table that exhibits pattern in respect to local odds ratios, these two graphical approaches can be applied for identifying pattern of odds ratios. These two graphical representations are the local odds ratios heat map square table plot and the three-dimensional local odds ratios heat map bar plot.

8.2.1 Heat Map Square Table

A heat map is a graphical representation of data where individual values contained in a matrix are represented as colors. Examining whether a table is symmetrical, a heat map serves as a good graphical tool. A heat map square table produces a square table that uses colors and the size of circles to represent the frequencies to examine symmetrical patterns of a table.

The following Tables 8.1 and 8.2 show a complete symmetry table, and a table that is deviated from complete symmetry, respectively. Figure 8.1 prints the heat map square tables for Table 8.1 and Fig. 8.2 for Table 8.2.

There are two sets of color schemes used in Fig. 8.1, one for the square and the other for the circle within the square. The purpose of having two sets of color

Table 8.1 Perfect symmetry table		C1	C2	C3
	R1	5	2	25
	R2	2	5	9
	R3	25	9	5

Table 8.2 Not a perfect symmetry Table

	C1	C2	C3
R1	5	2	15
R2	10	5	19
R3	25	3	5

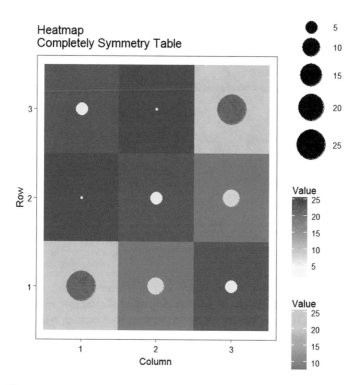

Fig. 8.1 Heat map square table—symmetry pattern

schemes in one table is to have contrast in colors for easier identification of
equivalency of their cell frequencies. We observe that the square for both cell (1,3)
and cell (3,1) has the same green color background and the circles in red color.
More importantly, the size of the two circles is exactly the same for both the cells,
indicating the equivalent of frequencies of the two cells. The bright blue background color of cell (1,2) and cell (2,1) and the small yellow dot within it also
indicating the equivalent of frequencies and their symmetrical position. The
yellow-colored circles for cell (2,3) and cell (3,2) display the same property of
symmetry with light purple background.

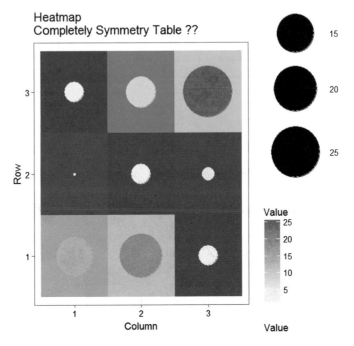

Fig. 8.2 Heat map square table—not in symmetry pattern

Unlike Fig. 8.1 above, the coloring of circles for Fig. 8.2 does not show equivalent symmetrical pattern as the coloring differ for the symmetrical cells. For instance, cell (1,3) has a big red circle with green background, whereas cell (3,1) has slightly smaller yellow circle with gray background. The contrary coloring between cell (1,2) and cell (2,1) and between cell (2,3) and cell (3,2) also highlighting the asymmetrical in frequencies.

The R syntax to generate the above two heat map square table plots is stated below.

```
HeatM_TotalS <- read.table(header=T, text='
Row Column Value
  1    1     25
  1    2     9
  1    3     5
  2    1     2
  2    2     5
  2    3     9
  3    1     5
  3    2     2
  3    3     25')
```

```
HeatM_TotalS2 <- read.table(header=T, text='
Row Column Value
  1    1     15
  1    2     19
  1    3     5
  2    1     2
  2    2     5
  2    3     3
  3    1     5
  3    2     10
  3    3     25')
library(ggplot2)
HP1 <- ggplot(HeatM_TotalS, aes(x = factor(Column), y = factor(Row))) +
       geom_tile(aes(fill = Value)) +
       scale_fill_continuous(low = "blue", high = "green")
HP1 <- HP1 + geom_point(aes(colour = Value, size =Value)) +
       scale_color_gradient(low = "yellow", high = "red") +
       scale_size(range = c(1, 15)) +
       theme_bw() +
       ggtitle("Heatmap\nCompletely Symmetry Table") +
       xlab("Column") +
       ylab("Row")
HP1
HP2 <- ggplot(HeatM_TotalS2, aes(x = factor(Column), y = factor(Row))) +
       geom_tile(aes(fill = Value)) +
       scale_fill_continuous(low = "blue", high = "green")
HP2 <- HP2 + geom_point(aes(colour = Value, size =Value)) +
       scale_color_gradient(low = "yellow", high = "red") +
       scale_size(range = c(1, 25)) +
       theme_bw() +
       ggtitle("Heatmap\nCompletely Symmetry Table ??") +
       xlab("Column") +
       ylab("Row")
HP2
library(gridExtra)
grid.arrange(HP1, HP2, ncol=2)
```

For both Tables 8.1 and 8.2, they are read in as data frames using the read.table() function. As the option "header = T" is specified, the first line of the data states the name of the variables. The header specifies three variables, row, column, and value, representing the row number, column number, and the frequencies of the row and column of the table. It is noted that the order of row and column specification in the read.table is different from the order of that in Tables 8.1 and 8.2. The order of Tables 8.1 and 8.2 starting from row 1 column 1, increase the column by 1 to row

1 column 2, proceed to row 1 column 3, and end up with row 3 and column 3. The order of input order, however, starting from cell (3,3), followed by cell (3,2), cell (3,1), cell (2,3), cell (2,2), and ending with cell (1,1). The reason for this reversed order between input in read.table() and the raw table is to ensure that the order for the graphical heat map square table plot is exactly the same as the order of table cell position. So, a direct comparison of table and graphical output is possible.

The ggplot() function is a graphical function from package ggplot2. Package ggplot2 is one of the commonly used R graphical packages. Package ggplot2 is an add-on package, so it has to be installed and loaded into the memory in order to use the ggplot function. The command library (ggplot2) read in the ggplot2 library. The first argument of the ggplot function is the name of the data frame. They are read in as HeatM_TotalS and HeatM_TotalS2 data frame, respectively, for Tables 8.1 and 8.2. The next argument is aes command that specifies the variables for x-axis and y-axis. The geom_tile (aes(fill = Value)) calls the output to fill with color from the variable Value. The scale_fill_continuous specifies the colors scheme that low frequencies to start with blue color and ends up with green color for high frequencies.

The geom_point commands specify the circle printed in the heat map. The variable Value specifies the color and size of circle. The scale_color_gradient command specifies the low frequencies starting with color yellow and high frequencies in red. The command scale_size specifies the size of the circle. The command theme_bw is the classic dark-on-light ggplot2 theme for non-data display. The ggtitle command prints the title, displayed on the top of the graph. The xlab() and ylab() specify the labels for x-axis and y-axis, respectively.

The library gridExtra controls the display of multiple graphs into one graphical output. By specifying the option ncol = 2 using grid.arrange() function, it places the two output graphs HP1 and HP2 side by side into two columns as one graphical plot.

8.2.2 Three-Dimensional Heat Map Bar Plot

The heat map square table transforms a doubly classified table into a two-dimensional plot represented in colors for the background and sizes of circle that map frequencies into colors. A three-dimensional heat map bar plot is another way of depicting a table by mapping it into a graphical three-dimensional space. As it is a three-dimensional graph, there are 3 axes. The x-axis and the y-axis represent the row and column categories, respectively, while the z-axis, the vertical axis, shows the frequencies of a doubly classified table with bars. The height follows a color scheme of bars. The color label indicating the height of the bar is printed on the right-hand side of the plot. The following syntax produce the three-dimensional heat map bar plot using package lattice cloud() function and package latticeExtra grid.arrange() function.

```
library(lattice)
library(latticeExtra)
TotalS <- matrix(c(5, 2, 8,
                   2, 5, 9,
                   8, 9, 5),
                   nrow=3)
rownames(TotalS) <- c(1,2,3)
colnames(TotalS) <- c(1,2,3)
TotalS2 <160;matrix(c( 5,   2,  25,
                       2,   5,   9,
                      15,  19,   5),
                     nrow=3)
rownames(TotalS2) <- c(1,2,3)
colnames(TotalS2) <160;c(1,2,3)
C1 <-
cloud(TotalS, panel.3d.cloud =160;panel.3dbars,
      xbase = 0.4, ybase = 0.4, zlim =160;c(0, max(TotalS)),
      scales =160;list(arrows =160;FALSE, just = "right"),
      xlab = "Row", ylab = "column", zlab="Frequency",
      main = "Cloud 3D Bars with Heatmap\nComplete Symmetry Table",
      col.facet = level.colors(TotalS, at = do.breaks(range(TotalS), 10),
                               col.regions = rainbow,
                               colors = TRUE),
      colorkey = list(col = rainbow, at = do.breaks(range(TotalS), 10)),
      screen = list(z = 40, x = -30))
C2 <-
cloud(TotalS2, panel.3d.cloud = panel.3dbars,
      xbase = 0.4, ybase = 0.4, zlim = c(0, max(TotalS2)),
      scales = list(arrows = FALSE, just = "right"),
      xlab = "Column", ylab = "Row", zlab="Frequency",
      main = "Cloud 3D Bars with Heatmap\nComplete Symmetry Table ??",
      col.facet = level.colors(TotalS2, at=do.breaks(range(TotalS2),10),
                               col.regions = rainbow,
                               colors = TRUE),
      colorkey = list(col = rainbow, at = do.breaks(range(TotalS2), 10)),
      screen = list(z = 40, x = -30))
grid.arrange(C1, C2, ncol=2)
```

The cloud function provides three-dimensional scatter plot and wireframe sur-
face plot. For the purpose of generating a graphical output to represent a table, a
three-dimensional bar plot with colors to represent the height of the bars is the main
aim. As colors are used for presenting the length of the bars that map to the
frequencies of table, it is referred to as a three-dimensional heat map bar plot.

Data have to specify as a matrix for inputting to cloud function. The rownames and colnames command assign the row and column number to the matrix. The first argument of cloud is the name of a matrix. panel.3d.cloud is one of the panel utilities available in the latticeExtra package. Specifying panel.3dbars produces three-dimensional bars. The command xbase and ybase specify the size of the bar. The default value is 1. This produces bar with no spaces between them. The lower the values assigned to xbase and ybase, the thinner is the bar chart. The length of the z-axis specifies zero as the base, and the maximum height is the maximum of all the frequencies of the table [xlim = c(0, max(TotalS))]. The labels for the 3 axes are specified by the xlab, ylab, and zlab options. The title of the graph is specified by the main option. The col.facet command specifies the graphical parameters for surfaces of the bars. The colorkey specifies whether a color key is to be drawn alongside the plot. The rainbow color scheme is used to produce the color of the graph. The label printed on the right-hand side of the plot shows the rainbow color in red on the bottom for low frequencies and pink on the top indicating those bars in high frequencies. The separation of the color is specified by the range of the table frequencies. The screen command gives the positioning of the bar plot in a three-dimensional space.

The left-hand side of Fig. 8.3, the 3-D heat map bar plot, shows the total symmetry in colors and length of the bars of the off-diagonal cells, whereas the diagonal bars are with the same color in green and with the same height. These characteristics in colors and bar lengths show that it is well suited for fitting a complete symmetry model. On the contrary, the right-hand side 3-D heat map bar plot does not show such symmetrical patterns and colors with the exception of the two red-colored bars and the three brown diagonal bars. This graph is thus not suited for a complete symmetry square table model.

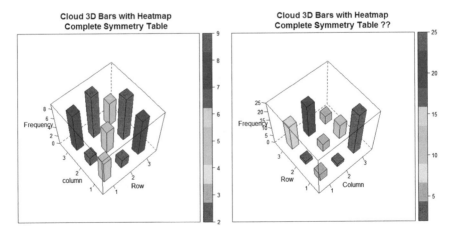

Fig. 8.3 Three-dimensional heat map bar plot—complete symmetry table

8.2.3   Local Odds Ratio Square Table Heat Map Plot

The two graphics in Sects. 8.2.1 and 8.2.2 apply to raw frequencies for identification of symmetrical patterns. The same principle can be applied to graph local odds ratios table. Table 8.3 tabulates the life satisfaction of Singaporean across two time period 2016 and 2017. The R syntax to draw local odds ratio square table heat map plot is shown below.

The following syntax reads in the data of the above table as a matrix and produces the local odds ratios in matrix.

```
QPData <- matrix(c(700, 130,  15,   15,
                   100, 330,  60,   30,
                    15,  60, 120,   70,
                    10,  20,  60, 650), nrow=4)
rownames(QPData) <- c(1,2,3,4)
colnames(QPData) <- c(1,2,3,4)
QPData
# Calculate Local Odds Ratios
a <- nrow(QPData)
LocalOdds<-matrix(data=NA,nrow=a-1,ncol=a-1)
for (i in 1:a-1) {
    for (j in 1:a-1) {
        LocalOdds[i,j] <- (QPData[i,j] * QPData[i+1,j+1]) / (QPData[i+1,j]
* QPData[i,j+1])
    }
}
rownames(LocalOdds) <- c(1,2,3)
colnames(LocalOdds) <- c(1,2,3)
LocalOdds
```

```
> QPData
    1   2   3   4
1 700 100  15  10
2 130 330  60  20
3  15  60 120  60
4  15  30  70 650
```

```
> LocalOdds
          1         2        3
1 17.769231  1.212121  0.50000
2  1.575758 11.000000  1.50000
3  0.500000  1.166667 18.57143
```

As function ggplot2 reads in long form of data, library reshape, melt() function is used to transform the matrix LocalOdds into a long form. The rest of the syntax is

Table 8.3 Life satisfaction of Singaporean, year 2016 × year 2017	Life satisfaction year 2016	Life satisfaction year 2017			
		1	2	3	4
	1	700	130	15	15
	2	100	330	60	30
	3	15	60	120	70
	4	10	20	60	650

the same as what have discussed. The local odds ratios square table heat map plot below shows the symmetrical pattern of the circles as well as square colors are in reference to the center of the table. The two red circles of cell (1,3) and cell (3,1) are in reference to the center of the plot. The cell (2,2) yellow dot is in symmetrical to the cell (3,2), and the small yellow dot of cell (1,1) and cell (3,3) are also in symmetrical to the center. So, it is very likely that a quasi point symmetry model suits well (Fig. 8.4).

```
library(reshape)
localOddsLong <- melt(LocalOdds)
library(ggplot2)
LO1 <- ggplot(localOddsLong , aes(x = factor(X1), y = factor(X2))) +
    geom_tile(aes(fill = value)) +
    scale_fill_continuous(low = "blue", high = "green") +
    geom_point(aes(colour = value, size =value)) +
    scale_color_gradient(low = "yellow", high = "red") +
    scale_size(range = c(1, 15)) +
    theme_bw() +
    ggtitle("Heatmap\nQuasi Point Symmetry Table\nLocal Odds Ratio") +
    xlab("Column") +
    ylab("Row")
LO1
```

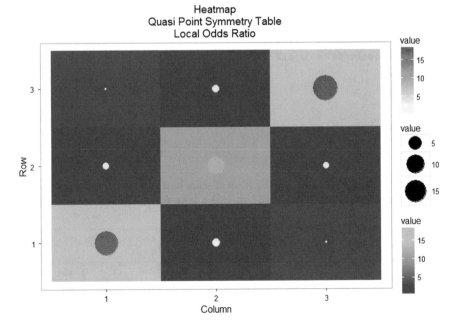

Fig. 8.4 Heat map—quasi-point symmetry table

8.2.4   Local Odds Ratio Square Table Three-dimensional Heat Map Bar Plot

The R syntax for producing a three-dimensional heat map bar plot for the local odds of Table 8.3 is stated below. Similarly, the symmetrical patterns of the bars in reference to the center of the bar plot are observed in this plot (Fig. 8.5).

```
library(latticeExtra)
cloud(LocalOdds, panel.3d.cloud = panel.3dbars,
        xbase = 0.4, ybase = 0.4, zlim = c(0, max(LocalOdds)),
        scales = list(arrows = FALSE, just = "right"),
        xlab = "Row", ylab = "column", zlab="Frequency",
        main = "Cloud 3D Bars with Heatmap\nQuasi Point Symmetry Table",
        col.facet  =  level.colors(LocalOdds,  at  =  do.breaks(range
(LocalOdds), 5),
                                    col.regions =160;rainbow,
                                    colors = TRUE),
 colorkey = list(col = rainbow, at = do.breaks(range(LocalOdds), 5)),
 screen = list(z = 40, x = -30))
```

Fig. 8.5 Three-dimensional bar chart heat map plot—quasi point symmetry table

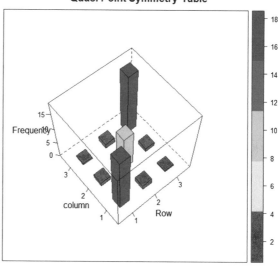
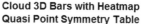

8.2.5   Package gplots Square Table Balloon Plot

Square table balloon plot is the fifth plot recommended for examining the association of cells of a doubly classified table. Package gplots function balloonplot produces the balloon plot. It plots a graphical matrix where each cell contains a dot whose size reflects the relative magnitude of the corresponding frequencies of a matrix/table. The following Table 8.4 shows the tabulation of assessments of two raters A and B.

The following gives the R syntax for producing the square table balloon plot of Table 8.3. The table is read in as a data frame called Rating using read.table() function specified with header. The library gplots function balloonplot generates the balloon plot. The data frame Rating is transformed three times to input to balloonplot function. First, it is transformed from a data frame into a matrix using as.matrix function, and second transformation from a matrix into a table using as.table function. It is further transposed using the t() function to input to the ballonplot function.

```
Rating <- read.table(header=T, text='
      RateA RateB RateC RateD RateE RateF
RateA  50     20     42     16     43     21
RateB  21     20     24     12     17     25
RateC  40     21     60     20     13     16
RateD  15     11     21     40     32     23
RateE  40     13     12     34     55     11
RateF  20     23     25     20     10     66')
library(gplots)
dt <- as.table(as.matrix(Rating))
balloonplot(t(dt), main ="Ballon Plot Rating\n Odds Symmetry Model 1",
            xlab ="", ylab="",
            label = FALSE, show.margins = FALSE)
```

Table 8.4 Rater A × Rater B

Rater A rating	Rater B rating					
	A	B	C	D	E	F
A	50	20	42	16	43	21
B	21	20	24	12	17	25
C	40	21	60	20	13	16
D	15	11	21	40	32	23
E	40	13	12	34	55	11
F	20	23	25	20	10	66

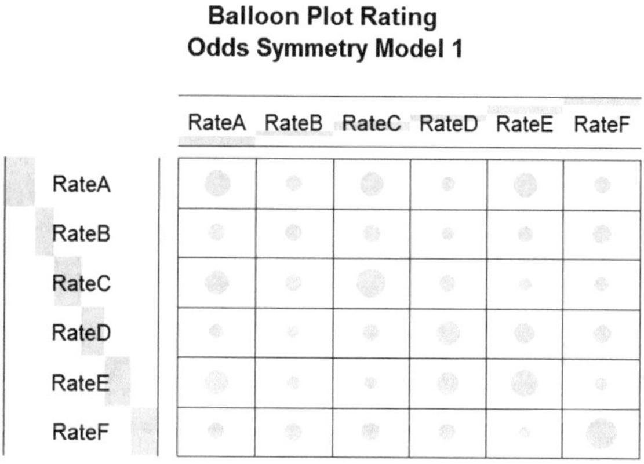

Fig. 8.6 Balloon plot—odds
symmetry model I

**Balloon Plot Rating
Odds Symmetry Model 1**

The balloon plot below shows that an odds symmetry model I is possible to fit the data. The pattern of size of circles for the first row and the first column is very similar. The second row from third column onwards and the second column from third row onwards also exhibit similar pattern in their size of circles. The symmetrical pattern is observed for the rest of the rows and columns. However, it is not that easy to see the pattern through naked eye without prior pointing out the pattern that follows the odds symmetry model I. There are limitations in graphical approach. Modeling using nonstandard log-linear is a preferred choice when the graphical approach fails (Fig. 8.6).

8.2.6   Mosaic Plot

Another way to represent doubly classified table in graphical form is the mosaic plot. A mosaic plot replaces absolute frequencies with relative frequencies represented counts as rectangular areas (Friendly 2000). The vertical and horizontal length of each rectangle is proportional to the proportion of the variables. The R way of producing a mosaic plot is to first create a table using the as.table() function and use it as input to the moasicplot() function to produce the mosaic plot, as shown in the syntax below (Fig. 8.7).

```
Totals <- as.table(matrix(c( 5, 2, 15,
                             2, 5,  9,
                            15, 9,  5),
                   nrow=3))
Totals <- as.table(Totals)
Totals
mosaicplot(Totals, main = "Mosaic Plot", color=2:4)
```

Fig. 8.7 Mosaic plot—
perfect symmetry table

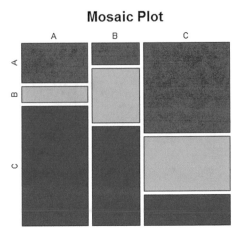

From the mosaic plot, we can observe that cell (1,1) in red, cell (2,2) in green, and cell (3,3) in blue having the same rectangular area. Similarly, cell (1,2) in red and cell (2,1) in green also have the same area space. But mosaic plot is not as direct as the other plots discussed earlier. This is the limitation of mosaic plot, as it is difficult to guess the equivalent of area space.

Exercises

8.1 Draw a heat map for the following Table 8.5 using ggplot2 package, ggplot() function. What do you observe from the plot?

8.2 Draw a three-dimensional bar chart for the following Table 8.6 using lattice package, cloud() function. What do you observe from the plot?

8.3 The following Table 8.7 is the cross-tabulation of 626 students expressed their self-efficacy level during two surveys carried out in year 2016 and 2017. Calculate the local odds ratio for the table. Draw a three-dimensional bar chart for the table and the local odds ratio table using lattice package, cloud() function. What do you observe from the plot?

Table 8.5 Happiness of
teachers 2015 and 2016

Happiness 2015	Happiness 2016					
	1	2	3	4	5	6
1	75	77	36	16	5	1
2	19	80	37	20	9	4
3	10	40	71	30	23	1
4	4	23	30	71	40	10
5	3	9	20	31	80	20
6	1	6	16	30	77	75

Table 8.6 Happiness of teachers 2015 and 2016

| Happiness 2015 | Happiness 2016 | | | | | |
	1	2	3	4	5	6
1	75	58	36	16	5	1
2	29	90	37	20	9	4
3	10	40	51	30	23	1
4	4	23	30	51	40	10
5	3	9	20	31	90	30
6	1	6	16	30	57	75

Table 8.7 Self-efficacy of students— year 2016 × year 2017

| Self-efficacy 2016 | Self-efficacy 2017 | | | |
	1	2	3	4
1	70	15	23	81
2	28	34	10	51
3	47	11	35	30
4	76	27	15	73

Reference

Friendly M. (2000). *Visualizing categorical data*. Cary, NC: SAS Institute.

Chapter 9
Creating Doubly Classified Models

9.1 Reverse Complete Symmetry Model

The complete symmetry model exhibits the symmetry in cells of off-diagonal as shown in the left-hand symbolic table (Fig. 9.1). Could the symmetrical pattern reverse? How about the diagonal cells that run from right top to left bottom instead of from the left top down to the right bottom, as shown in the right-hand side table of Fig. 9.1? How could we model it? Can a new complete symmetry of the probabilities that is reverse that of complete symmetry model be set up using nonstandard log-linear model?

The specification for this new model is quite straightforward. This is done by specification a new factor variable. We shall call this model the reverse complete symmetry model (RS) as it is reversed in symmetry that of the complete symmetry model (S). The nonstandard log-linear specification of the reversed complete symmetry is as follows.

$$S = \begin{bmatrix} 1 & 2 & 3 & 4 \\ 2 & 5 & 6 & 7 \\ 3 & 6 & 8 & 9 \\ 4 & 7 & 9 & 10 \end{bmatrix}$$

$$\ln(m_{ij}) = \mu + \lambda_{ij}^{RS}$$

where $RS = \begin{bmatrix} 4 & 3 & 2 & 1 \\ 7 & 6 & 5 & 2 \\ 9 & 8 & 6 & 3 \\ 10 & 9 & 7 & 4 \end{bmatrix}$ for a 4 × 4 table.

We observe that the numeric specification of the factor RS of reversed complete symmetry is just the reverse of the complete symmetry factor S. In short, the specification of equivalent in probability of cells is to assign the same numeric

© Springer Nature Singapore Pte Ltd. 2017
T.K. Tan, *Doubly Classified Model with R*,
https://doi.org/10.1007/978-981-10-6995-6_9

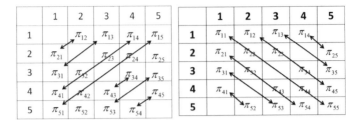

Fig. 9.1 Complete symmetry and reverse complete symmetry model

Table 9.1 Depression inventory—Item 1 × Item 2

Item 1	Item 2			
	1	2	3	4
1	97	47	55	33
2	40	70	32	57
3	52	35	72	43
4	38	50	42	96

value to the factor and use nonstandard log-linear to generate the model with the specified factor. As long as we specify a new factor structure, we get a new doubly classified model. In general, by specifying the factor value to be the same, we are stipulating the equality of the parameter. For instance, cell (1,1) and cell (4,4) are specified with the same factor value 4, the estimated values for cell (1,1) is equal to cell (4,4), and the estimated values are simple the average of the two values, using ML estimation. But for complete symmetry model, it is cell (1,4) and cell (4,1) are specified as in equality. As such, the factor specification of the complete symmetry is in reverse of that of the reverse complete symmetry model. As far as we hypothesize the probabilities of the two cells are supposedly to be equal, based on theoretical reasoning, we can build a new doubly classified model according to our specification.

Table 9.1 gives an example to generate a reverse complete symmetry model. The table records 859 patients with two items to measure depression level that ranges from 1 to 4 being 1 representing low depression level to 4 indicating high depression level.

The R syntax to generate the reverse complete symmetry model is given below.

```
ComRS <- c(97, 47, 55, 33,
           40, 70, 32, 57,
           52, 35, 72, 43,
           38, 50, 42, 96)
# --------------------------- #
# Reversed Complete Symmetry Model #
# --------------------------- #
```

```
rs<-c( 4, 3, 2, 1,
       7, 6, 5, 2,
       9, 8, 6, 3,
       10, 9, 7, 4)
rs<-as.factor(rs)
r<-gl(4,4)
c<-gl(4,1,length=16)
RS<-glm(ComRS~rs,family=poisson)
summary(RS)
model.summary(RS)
```

The estimates for the reverse complete symmetry model are shown below where $\hat{m}_{ij} = (n_{ij} + n_{i+k,j+k})/2$, and $k = (a+1) - (i+j)$. The degrees of freedom for reverse complete symmetry is the same as complete symmetry, $a(a-1)/2$, since the number of factors used to model them are the same.

The fitted values for reverse complete symmetry model are stated below.

$$\hat{m}_{ij} = \begin{bmatrix} 96.5 & 45 & 56 & 33 \\ 41 & 71 & 32 & 56 \\ 51 & 35 & 71 & 45 \\ 38 & 51 & 41 & 96.5 \end{bmatrix}$$

9.2   Parallel Diagonal Symmetry Model

The reverse complete symmetry model discussed in the last section demonstrates a new doubly classified model that can be easily set up by specifying a new factor structure. This section illustrates another new doubly classified model called parallel diagonal symmetry model, with the following symbolic table in mind. The symbolic table indicates that those cells with the same color enclosed are with the same probabilities. For instance, $n_{12} = n_{21}$, $n_{23} = n_{32}$, $n_{34} = n_{43}$, $n_{45} = n_{54}$ and these cells are with the same estimated values (Fig. 9.2).

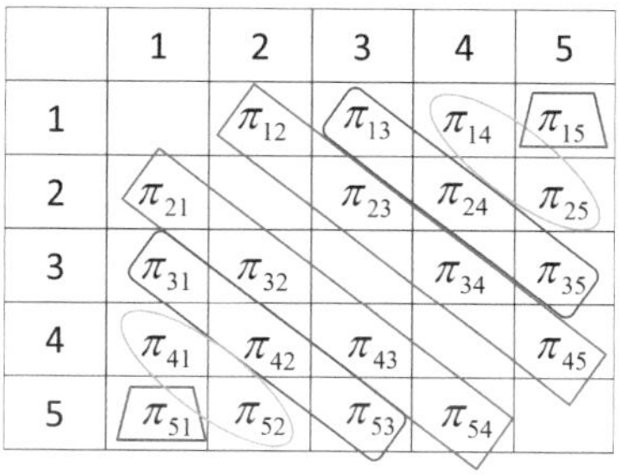

Fig. 9.2 Parallel diagonal symmetry model

The nonstandard model of parallel diagonal symmetry model is specified as follows:

$$\ln(m_{ij}) = \mu + \lambda_{ij}^{PD}$$

$$PD = \begin{bmatrix} 1 & 6 & 7 & 8 & 9 \\ 6 & 2 & 6 & 7 & 8 \\ 7 & 6 & 3 & 6 & 7 \\ 8 & 7 & 6 & 4 & 6 \\ 9 & 8 & 7 & 6 & 5 \end{bmatrix}$$

The following table tabulates the rating of rater A and rater B for 870 students (Table 9.2).

The R syntax to generate the parallel diagonal symmetry model is given below.

```
Rater <- c(230, 12, 32, 11,  8,
            13, 56, 11, 23, 17,
            29, 12, 77, 12, 15,
            13, 21, 15, 88, 19,
             6, 14, 22, 15, 99)
PD <- c( 1, 6, 7, 8, 9,
         6, 2, 6, 7, 8,
         7, 6, 3, 6, 7,
         8, 7, 6, 4, 6,
         9, 8, 7, 6, 5)
PD <- as.factor(PD)
PDS <-glm(Rater~PD,family=poisson)
summary(PDS)
model.summary(PDS)
```

Table 9.2 Rater A × Rater B

Rater B	Rater A				
	1	2	3	4	5
1	230	12	32	11	8
2	13	56	11	23	17
3	29	12	77	12	15
4	13	21	15	88	19
5	6	14	22	15	99

The estimates for the parallel diagonal symmetry model are shown below.

$$
\hat{m}_{ij} = \begin{bmatrix}
230 & 13.625 & 23.667 & 13.75 & 7 \\
13.625 & 56 & 13.625 & 23.667 & 14.75 \\
23.667 & 13.625 & 77 & 13.625 & 23.667 \\
13.75 & 23.667 & 13.625 & 88 & 13.625 \\
7 & 13.75 & 23.667 & 13.625 & 99
\end{bmatrix}
$$

9.3 Quasi-Symmetry with N Degree Models

The nonstandard log-linear formulation of complete symmetry model is to specify a factor S as shown in the previous section. By adding two additional factors, R and C, a quasi-symmetry model (QS) is specified. The nonstandard log-linear specification of QS and the three factors are stated below. This main characteristic of QS is it exhibits symmetry in odds ratios.

$$
\ln(\hat{m}_{ij}) = \mu + \lambda_i^R + \lambda_j^C + \lambda_{ij}^S
$$

$$
S = \begin{bmatrix}
1 & 2 & 3 & 4 \\
2 & 5 & 6 & 7 \\
3 & 6 & 8 & 9 \\
4 & 7 & 9 & 10
\end{bmatrix}, R = \begin{bmatrix}
1 & 2 & 3 & 4 \\
1 & 2 & 3 & 4 \\
1 & 2 & 3 & 4 \\
1 & 2 & 3 & 4
\end{bmatrix}, C = \begin{bmatrix}
1 & 1 & 1 & 1 \\
2 & 2 & 2 & 2 \\
3 & 3 & 3 & 3 \\
4 & 4 & 4 & 4
\end{bmatrix}
$$

Similarly, the nonstandard log-linear model of quasi-point symmetry model is similar to that of the quasi-symmetry model with two additional factors R and C but instead of factor S, factor P is specified for quasi-point symmetry model.

$$
\ln(\hat{m}_{ij}) = \mu + \lambda_i^R + \lambda_j^C + \lambda_{ij}^P
$$

$$
S = \begin{bmatrix}
1 & 2 & 3 & 4 \\
2 & 5 & 6 & 7 \\
3 & 6 & 8 & 9 \\
4 & 7 & 9 & 10
\end{bmatrix}, R = \begin{bmatrix}
1 & 2 & 3 & 4 \\
1 & 2 & 3 & 4 \\
1 & 2 & 3 & 4 \\
1 & 2 & 3 & 4
\end{bmatrix}, P = \begin{bmatrix}
1 & 3 & 4 & 5 \\
8 & 2 & 6 & 7 \\
7 & 6 & 2 & 8 \\
5 & 4 & 3 & 1
\end{bmatrix}
$$

The S and R factors set the odds ratios to be symmetry. The adding of C factor specifies the off-diagonal to be symmetrical in probabilities for the quasi-symmetry model , whereas the adding of P factor specifies the point symmetry to the center of the table for quasi-point symmetry model. The S and R factors are the two crucial factors that generate the symmetry of odds ratio and adjusts according to the specification of the third factor.

The additional of a fourth factor on top of the R, C, and S generates restriction on the odds ratios. Quasi-conditional symmetry model displays symmetry in odds ratios but with restriction to the symmetry of odds ratios to the off-diagonal that are two positions away from the diagonal. Those cells that are one position away from the diagonal has a constant γ ratio between upper and lower diagonal cells. The nonstandard log-linear model specification is to include R, C, S, and CS factors as shown below.

$$\ln\left(\hat{m}_{ij}\right) = \mu + \lambda_i^R + \lambda_j^C + \lambda_{ij}^S + \eta^{CS}$$

Now, we turn to generate a series of model with symmetry in odds ratios but with restrictions. How about we restrict the symmetry in odds ratios to only those closer to the diagonal? The above principal of adding a fourth factor applies. Add in a fourth factor that specifies the condition for the restriction will generate a different restriction of a new model.

Now, we proceed to generate a series of new models named as quasi-symmetry of n degree, denoted as QS(n) (Tan and Sheng 2015). These models are a series of general forms of the quasi-symmetry model (QS), further relaxing the quasi-symmetry model (QS) assumptions on the odds ratios. While the QS model exhibits the symmetry in odds ratios from the diagonal cells of a doubly classified table, the QS(1), QS model with degree 1, restricts those cells closest to the diagonal. When the number of cells in symmetry is further restricted to those cells that are further away from the diagonal, a series of QS models with symmetry degree n are specified, referred to as the QS(n) model. The difference between QS and QS(n) is the additional specification of the QS(n) factors.

The main characteristics of QS(n) is that $\hat{\theta}_{ij} = \hat{\theta}_{ji}$ for $|i - j| \leq n$. The symbolic tables below show the symmetrical pattern of odds ratios for QS(n). Given a 10×10 doubly classified table, a 9×9 sub square table that summarizes the local odds ratios is obtained as those shown in Fig. 9.3. Figure 9.3 shows a QS(1) model in which those local odds ratios have the property of symmetry, whereas those blank cells are not in symmetry. For instance, $\theta_{12} = \theta_{21}$ and $\theta_{23} = \theta_{32}$ are cells that are one position away from the diagonal having equality in their odds ratios. For those local odds ratios not stated in Fig. 9.3, they do not possess the property of equality in their local odds ratio, e.g., $\theta_{13} \neq \theta_{31}$. While Fig. 9.3 of QS(1) model

Fig. 9.3 QS(1) model

	1	2	3	4	5	6	7	8	9
1		θ_{12}							
2	θ_{21}		θ_{23}						
3		θ_{32}		θ_{34}					
4			θ_{43}		θ_{45}				
5				θ_{54}		θ_{56}			
6					θ_{65}		θ_{67}		
7						θ_{76}		θ_{78}	
8							θ_{87}		θ_{89}
9								θ_{98}	

QS(1) $\theta_{12} = \theta_{21}, \ldots, \theta_{89} = \theta_{98}$

restricts to the symmetry of odds ratios on one position off-diagonal, QS(2) further restricts the symmetry of odds ratios to the next position off the diagonal as shown in Fig. 9.4. For QS(2), not only $\theta_{12} = \theta_{21}$, $\theta_{23} = \theta_{32}$, and those that are one off the diagonal are in symmetry, those odds ratios that are two positions off the diagonal are also equivalent in their odds ratios such as $\theta_{13} = \theta_{31}$, and $\theta_{24} = \theta_{42}$. When the degree of symmetry n goes up, more cells will be restricted to having equality in their odds ratios. For QS(3), it restricts to cells up to three positions away from the diagonal. The same procedure applies up to QS(7) where the symmetry of odds ratios extends to the seventh position. If the eighth position of the cells is also in symmetry, the QS(8) is equivalent to the QS model since all the cells are in symmetry (Fig. 9.5).

Fig. 9.4 QS(2) model

	1	2	3	4	5	6	7	8	9
1		θ_{12}	θ_{13}						
2	θ_{21}		θ_{23}	θ_{24}					
3	θ_{31}	θ_{32}		θ_{34}	θ_{35}				
4		θ_{42}	θ_{43}		θ_{45}	θ_{46}			
5			θ_{53}	θ_{54}		θ_{56}	θ_{57}		
6				θ_{64}	θ_{65}		θ_{67}	θ_{68}	
7					θ_{75}	θ_{76}		θ_{78}	θ_{79}
8						θ_{86}	θ_{87}		θ_{89}
9							θ_{97}	θ_{98}	

QS(2) $\theta_{12} = \theta_{21},\ldots, \theta_{89} = \theta_{98}$ **and** $\theta_{13} = \theta_{31},\ldots, \theta_{79} = \theta_{97}$

Fig. 9.5 QS(3) model

	1	2	3	4	5	6	7	8	9
1		θ_{12}	θ_{13}	θ_{14}					
2	θ_{21}		θ_{23}	θ_{24}	θ_{25}				
3	θ_{31}	θ_{32}		θ_{34}	θ_{35}	θ_{36}			
4	θ_{41}	θ_{42}	θ_{43}		θ_{45}	θ_{46}	θ_{47}		
5		θ_{52}	θ_{53}	θ_{54}		θ_{56}	θ_{57}	θ_{58}	
6			θ_{63}	θ_{64}	θ_{65}		θ_{67}	θ_{68}	θ_{69}
7				θ_{74}	θ_{75}	θ_{76}		θ_{78}	θ_{79}
8					θ_{85}	θ_{86}	θ_{87}		θ_{89}
9						θ_{96}	θ_{97}	θ_{98}	

QS(3) $\theta_{12} = \theta_{21},\ldots, \theta_{89} = \theta_{98}$, $\theta_{13} = \theta_{31},\ldots, \theta_{79} = \theta_{97}$
and $\theta_{14} = \theta_{41},\ldots, \theta_{69} = \theta_{96}$

The following shows the nonstandard log-linear model specification of QS(n) model.

$$\ln\left(\hat{m}_{ij}\right) = \mu + \lambda_i^R + \lambda_j^C + \lambda_{ij}^S + \lambda_{ij}^{QS(n)}$$

$$df = \frac{(a-1)(a-2)}{2} - (a-n-2)$$

where the QS(n) factor is defined as follows:

$$OS(n)_{ij} = \begin{cases} 1 & i = j \\ k+1 & i > j \\ k+1 & i < j \,\&\, k \le n \\ n+2 & i < j \,\&\, k > n \end{cases}$$

$$k = j - i$$

The factors of QS(1) for 10×10 table are as follows:

$$QS1 = \begin{bmatrix} 1 & 2 & 3 & 3 & 3 & 3 & 3 & 3 & 3 & 3 \\ 2 & 1 & 2 & 3 & 3 & 3 & 3 & 3 & 3 & 3 \\ 3 & 2 & 1 & 2 & 3 & 3 & 3 & 3 & 3 & 3 \\ 4 & 3 & 2 & 1 & 2 & 3 & 3 & 3 & 3 & 3 \\ 5 & 4 & 3 & 2 & 1 & 2 & 3 & 3 & 3 & 3 \\ 6 & 5 & 4 & 3 & 2 & 1 & 2 & 3 & 3 & 3 \\ 7 & 6 & 5 & 4 & 3 & 2 & 1 & 2 & 3 & 3 \\ 8 & 7 & 6 & 5 & 4 & 3 & 2 & 1 & 2 & 3 \\ 9 & 8 & 7 & 6 & 5 & 4 & 3 & 2 & 1 & 2 \\ 10 & 9 & 8 & 7 & 6 & 5 & 4 & 3 & 2 & 1 \end{bmatrix}$$

For QS(2), the factor for QS(2) is as follows:

$$QS2 = \begin{bmatrix} 1 & 2 & 3 & 4 & 4 & 4 & 4 & 4 & 4 & 4 \\ 2 & 1 & 2 & 3 & 4 & 4 & 4 & 4 & 4 & 4 \\ 3 & 2 & 1 & 2 & 3 & 4 & 4 & 4 & 4 & 4 \\ 4 & 3 & 2 & 1 & 2 & 3 & 4 & 4 & 4 & 4 \\ 5 & 4 & 3 & 2 & 1 & 2 & 3 & 4 & 4 & 4 \\ 6 & 5 & 4 & 3 & 2 & 1 & 2 & 3 & 4 & 4 \\ 7 & 6 & 5 & 4 & 3 & 2 & 1 & 2 & 3 & 4 \\ 8 & 7 & 6 & 5 & 4 & 3 & 2 & 1 & 2 & 3 \\ 9 & 8 & 7 & 6 & 5 & 4 & 3 & 2 & 1 & 2 \\ 10 & 9 & 8 & 7 & 6 & 5 & 4 & 3 & 2 & 1 \end{bmatrix},$$

For QS(3),

$$QS3 = \begin{bmatrix} 1 & 2 & 3 & 4 & 5 & 5 & 5 & 5 & 5 & 5 \\ 2 & 1 & 2 & 3 & 4 & 5 & 5 & 5 & 5 & 5 \\ 3 & 2 & 1 & 2 & 3 & 4 & 5 & 5 & 5 & 5 \\ 4 & 3 & 2 & 1 & 2 & 3 & 4 & 5 & 5 & 5 \\ 5 & 4 & 3 & 2 & 1 & 2 & 3 & 4 & 5 & 5 \\ 6 & 5 & 4 & 3 & 2 & 1 & 2 & 3 & 4 & 5 \\ 7 & 6 & 5 & 4 & 3 & 2 & 1 & 2 & 3 & 4 \\ 8 & 7 & 6 & 5 & 4 & 3 & 2 & 1 & 2 & 3 \\ 9 & 8 & 7 & 6 & 5 & 4 & 3 & 2 & 1 & 2 \\ 10 & 9 & 8 & 7 & 6 & 5 & 4 & 3 & 2 & 1 \end{bmatrix},$$

and so on.

Table 9.3 tabulates two indicators of a skills inventory, coded from 1 to 10 being 1 is the lowest for the indicator and 10 highest for the indicator.

The r, c, and s factor for 10×10 factor are as follows:

$$r = \begin{bmatrix} 1 & 1 & 1 & 1 & 1 & 1 & 1 & 1 & 1 & 1 \\ 2 & 2 & 2 & 2 & 2 & 2 & 2 & 2 & 2 & 2 \\ 3 & 3 & 3 & 3 & 3 & 3 & 3 & 3 & 3 & 3 \\ 4 & 4 & 4 & 4 & 4 & 4 & 4 & 4 & 4 & 4 \\ 5 & 5 & 5 & 5 & 5 & 5 & 5 & 5 & 5 & 5 \\ 6 & 6 & 6 & 6 & 6 & 6 & 6 & 6 & 6 & 6 \\ 7 & 7 & 7 & 7 & 7 & 7 & 7 & 7 & 7 & 7 \\ 8 & 8 & 8 & 8 & 8 & 8 & 8 & 8 & 8 & 8 \\ 9 & 9 & 9 & 9 & 9 & 9 & 9 & 9 & 9 & 9 \\ 10 & 10 & 10 & 10 & 10 & 10 & 10 & 10 & 10 & 10 \end{bmatrix}$$

$$c = \begin{bmatrix} 1 & 2 & 3 & 4 & 5 & 6 & 7 & 8 & 9 & 10 \\ 1 & 2 & 3 & 4 & 5 & 6 & 7 & 8 & 9 & 10 \\ 1 & 2 & 3 & 4 & 5 & 6 & 7 & 8 & 9 & 10 \\ 1 & 2 & 3 & 4 & 5 & 6 & 7 & 8 & 9 & 10 \\ 1 & 2 & 3 & 4 & 5 & 6 & 7 & 8 & 9 & 10 \\ 1 & 2 & 3 & 4 & 5 & 6 & 7 & 8 & 9 & 10 \\ 1 & 2 & 3 & 4 & 5 & 6 & 7 & 8 & 9 & 10 \\ 1 & 2 & 3 & 4 & 5 & 6 & 7 & 8 & 9 & 10 \\ 1 & 2 & 3 & 4 & 5 & 6 & 7 & 8 & 9 & 10 \\ 1 & 2 & 3 & 4 & 5 & 6 & 7 & 8 & 9 & 10 \end{bmatrix}$$

$$s = \begin{bmatrix} 1 & 2 & 3 & 4 & 5 & 6 & 7 & 8 & 9 & 10 \\ 2 & 11 & 12 & 13 & 14 & 15 & 16 & 17 & 18 & 19 \\ 3 & 12 & 20 & 21 & 22 & 23 & 24 & 25 & 26 & 27 \\ 4 & 13 & 21 & 28 & 29 & 30 & 31 & 32 & 33 & 34 \\ 5 & 14 & 22 & 29 & 35 & 36 & 37 & 38 & 39 & 40 \\ 6 & 15 & 23 & 30 & 36 & 37 & 41 & 42 & 43 & 44 \\ 7 & 16 & 24 & 31 & 37 & 42 & 46 & 47 & 48 & 49 \\ 8 & 17 & 25 & 32 & 38 & 43 & 47 & 50 & 51 & 52 \\ 9 & 18 & 26 & 33 & 39 & 44 & 48 & 51 & 53 & 54 \\ 10 & 19 & 27 & 34 & 40 & 45 & 49 & 52 & 54 & 55 \end{bmatrix}$$

The R syntax for fitting QS(1) to QS(7) is as follows:

```
Freq <- c(
   31,  8, 20, 29, 25, 30, 50, 22, 21, 42,
    2,  4,  4,  6,  3,  4,  4,  1,  3,  7,
    5,  4,  5,  6,  2,  9,  9,  7,  6,  6,
    6,  4,  4,  8, 18, 28, 32, 20, 12, 21,
    2,  1,  1, 11, 14,  5, 13,  7,  8,  5,
    3,  1,  5, 21,  7, 14, 27, 11, 11, 14,
    4,  2,  3, 30, 22, 36, 76, 23, 27, 37,
    2,  1,  4, 12, 11, 17, 26, 13, 18, 18,
    1,  1,  2, 12, 14, 23, 46, 27, 30, 38,
    3,  2,  4, 21, 19, 33, 114, 54, 75, 203)
r<-gl(10,10)
c<-gl(10,1,length=100)
s<-c( 1, 2, 3, 4, 5, 6, 7, 8, 9, 10,
      2,11,12,13,14,15,16,17,18,19,
      3,12,20,21,22,23,24,25,26,27,
```

Table 9.3 Literacy × Leadership skills

		Leadership									
		1	2	3	4	5	6	7	8	9	10
Literacy	1	31	8	20	29	25	30	50	22	21	42
	2	2	4	4	6	3	4	4	1	3	7
	3	5	4	5	6	2	9	9	7	6	6
	4	6	4	4	8	18	28	32	20	12	21
	5	2	1	1	11	14	5	13	7	8	5
	6	3	1	5	21	7	14	27	11	11	14
	7	4	2	3	30	22	36	76	23	27	37
	8	2	1	4	12	11	17	26	13	18	18
	9	1	1	2	12	14	23	46	27	30	38
	10	3	2	4	21	19	33	114	54	75	203

```
     4,13,21,28,29,30,31,32,33,34,
     5,14,22,29,35,36,37,38,39,40,
     6,15,23,30,36,41,42,43,44,45,
     7,16,24,31,37,42,46,47,48,49,
     8,17,25,32,38,43,47,50,51,52,
     9,18,26,33,39,44,48,51,53,54,
    10,19,27,34,40,45,49,52,54,55)
s<-as.factor(s)
qs1<-c( 1, 2, 3, 3, 3, 3, 3, 3, 3, 3,
        2, 1, 2, 3, 3, 3, 3, 3, 3, 3,
        3, 2, 1, 2, 3, 3, 3, 3, 3, 3,
        4, 3, 2, 1, 2, 3, 3, 3, 3, 3,
        5, 4, 3, 2, 1, 2, 3, 3, 3, 3,
        6, 5, 4, 3, 2, 1, 2, 3, 3, 3,
        7, 6, 5, 4, 3, 2, 1, 2, 3, 3,
        8, 7, 6, 5, 4, 3, 2, 1, 2, 3,
        9, 8, 7, 6, 5, 4, 3, 2, 1, 2,
       10, 9, 8, 7, 6, 5, 4, 3, 2, 1)
qs1<-as.factor(qs1)
qs2<-c( 1, 2, 3, 4, 4, 4, 4, 4, 4, 4,
        2, 1, 2, 3, 4, 4, 4, 4, 4, 4,
        3, 2, 1, 2, 3, 4, 4, 4, 4, 4,
        4, 3, 2, 1, 2, 3, 4, 4, 4, 4,
        5, 4, 3, 2, 1, 2, 3, 4, 4, 4,
        6, 5, 4, 3, 2, 1, 2, 3, 4, 4,
        7, 6, 5, 4, 3, 2, 1, 2, 3, 4,
        8, 7, 6, 5, 4, 3, 2, 1, 2, 3,
        9, 8, 7, 6, 5, 4, 3, 2, 1, 2,
       10, 9, 8, 7, 6, 5, 4, 3, 2, 1)
qs2<-as.factor(qs2)
qs3<-c( 1, 2, 3, 4, 5, 5, 5, 5, 5, 5,
        2, 1, 2, 3, 4, 5, 5, 5, 5, 5,
        3, 2, 1, 2, 3, 4, 5, 5, 5, 5,
        4, 3, 2, 1, 2, 3, 4, 5, 5, 5,
        5, 4, 3, 2, 1, 2, 3, 4, 5, 5,
        6, 5, 4, 3, 2, 1, 2, 3, 4, 5,
        7, 6, 5, 4, 3, 2, 1, 2, 3, 4,
        8, 7, 6, 5, 4, 3, 2, 1, 2, 3,
        9, 8, 7, 6, 5, 4, 3, 2, 1, 2,
       10, 9, 8, 7, 6, 5, 4, 3, 2, 1)
qs3<-as.factor(qs3)
qs4<-c( 1, 2, 3, 4, 5, 6, 6, 6, 6, 6,
        2, 1, 2, 3, 4, 5, 6, 6, 6, 6,
        3, 2, 1, 2, 3, 4, 5, 6, 6, 6,
        4, 3, 2, 1, 2, 3, 4, 5, 6, 6,
```

```
             5, 4, 3, 2, 1, 2, 3, 4, 5, 6,
             6, 5, 4, 3, 2, 1, 2, 3, 4, 5,
             7, 6, 5, 4, 3, 2, 1, 2, 3, 4,
             8, 7, 6, 5, 4, 3, 2, 1, 2, 3,
             9, 8, 7, 6, 5, 4, 3, 2, 1, 2,
            10, 9, 8, 7, 6, 5, 4, 3, 2, 1)
qs4<-as.factor(qs4)
qs5<-c( 1, 2, 3, 4, 5, 6, 7, 7, 7, 7,
             2, 1, 2, 3, 4, 5, 6, 7, 7, 7,
             3, 2, 1, 2, 3, 4, 5, 6, 7, 7,
             4, 3, 2, 1, 2, 3, 4, 5, 6, 7,
             5, 4, 3, 2, 1, 2, 3, 4, 5, 6,
             6, 5, 4, 3, 2, 1, 2, 3, 4, 5,
             7, 6, 5, 4, 3, 2, 1, 2, 3, 4,
             8, 7, 6, 5, 4, 3, 2, 1, 2, 3,
             9, 8, 7, 6, 5, 4, 3, 2, 1, 2,
            10, 9, 8, 7, 6, 5, 4, 3, 2, 1)
qs5<-as.factor(qs5)
qs6 < -c( 1, 2, 3, 4, 5, 6, 7, 8, 8, 8,
             2, 1, 2, 3, 4, 5, 6, 7, 8, 8,
             3, 2, 1, 2, 3, 4, 5, 6, 7, 8,
             4, 3, 2, 1, 2, 3, 4, 5, 6, 7,
             5, 4, 3, 2, 1, 2, 3, 4, 5, 6,
             6, 5, 4, 3, 2, 1, 2, 3, 4, 5,
             7, 6, 5, 4, 3, 2, 1, 2, 3, 4,
             8, 7, 6, 5, 4, 3, 2, 1, 2, 3,
             9, 8, 7, 6, 5, 4, 3, 2, 1, 2,
            10, 9, 8, 7, 6, 5, 4, 3, 2, 1)
qs6<-as.factor(qs6)
qs7<-c( 1, 2, 3, 4, 5, 6, 7, 8, 9,10,
             2, 1, 2, 3, 4, 5, 6, 7, 8, 9,
             3, 2, 1, 2, 3, 4, 5, 6, 7, 8,
             4, 3, 2, 1, 2, 3, 4, 5, 6, 7,
             5, 4, 3, 2, 1, 2, 3, 4, 5, 6,
             6, 5, 4, 3, 2, 1, 2, 3, 4, 5,
             7, 6, 5, 4, 3, 2, 1, 2, 3, 4,
             8, 7, 6, 5, 4, 3, 2, 1, 2, 3,
             9, 8, 7, 6, 5, 4, 3, 2, 1, 2,
            10, 9, 8, 7, 6, 5, 4, 3, 2, 1)
qs7<-as.factor(qs7)
# Quasi Symmetry Model
QS<-glm(Freq~r+c+s,family=poisson)
summary(QS)
model.summary(QS)
# Quasi Symmetry Model with Degree 1
```

```
QS1<-glm(Freq~r+c+s+qs1,family=poisson)
summary(QS1)
model.summary(QS1)
# Quasi Symmetry Model with Degree 2
QS2<-glm(LitPhy~r+c+s+qs2,family=poisson)
summary(QS2)
model.summary(QS2)
# Quasi Symmetry Model with Degree 3
QS3<-glm(LitPhy~r+c+s+qs3,family=poisson)
summary(QS3)
model.summary(QS3)
# Quasi Symmetry Model with Degree 4
QS4<-glm(LitPhy~r+c+s+qs4,family=poisson)
summary(QS4)
model.summary(QS4)
# Quasi Symmetry Model with Degree 5
QS5<-glm(LitPhy~r+c+s+qs5,family=poisson)
summary(QS5)
model.summary(QS5)
# Quasi Symmetry Model with Degree 6
QS6<-glm(LitPhy~r+c+s+qs6,family=poisson)
summary(QS6)
model.summary(QS6)
# Quasi Symmetry Model with Degree 7
QS7<-glm(LitPhy~r+c+s+qs7,family=poisson)
summary(QS7)
model.summary(QS7)
```

Tables 9.4, 9.5, and 9.6 show the odds ratios table for QS, QS(1), and QS(2), respectively. Table 9.4 shows total symmetry in odds ratios for a QS model. Those cells that are one position away from the diagonal of Table 9.5 have the same estimated odds ratio. For instance, cell (1,2) and cell(2,1) have the same estimated odds ratio of 0.4, cell (2,3) and cell (3,2) have the same estimated odds ratio of 0.6, and so on. Those cells that are two positions away from the diagonal are not in symmetry. For instance, the estimated odds ratio of cell (1,9) is 1.1 which is different from that of cell (9,1) of 0.6, the estimated odds ratio of cell (2,9) is 0.4 which is different from cell (9,2) of 0.8. Table 9.6 shows the output for QS(2) model which displays symmetry in odds ratios for those cells that are one and two positions away from the diagonal. In short, the symmetry in odds ratios that takes place only to those cells one step away from the diagonal for a QS(1) model, both one and two steps away from the diagonal for QS(2) model and so on. In general, a QS(n) model specifies those cells are symmetry in odds ratios up to n steps away from diagonal.

Table 9.4 Fitted odds ratio—QS

		Literacy								
		1	2	3	4	5	6	7	8	9
Literacy	1	8.0	0.4	1.0	0.6	1.0	0.7	0.7	2.1	1.0
	2	0.4	1.3	0.8	0.8	3.7	0.7	2.8	0.4	0.5
	3	1.0	0.8	1.7	6.6	0.4	1.5	0.6	1.0	1.3
	4	0.6	0.8	6.6	0.6	0.2	2.3	1.0	1.5	0.6
	5	1.0	3.7	0.4	0.2	5.5	0.8	0.9	1.0	1.3
	6	0.7	0.7	1.5	2.3	0.8	1.1	0.7	1.2	1.5
	7	0.7	2.7	0.6	1.0	0.9	0.7	1.6	1.2	0.8
	8	2.0	0.4	1.0	1.5	1.0	1.2	1.2	0.8	1.2
	9	1.0	0.5	1.3	0.6	1.3	1.5	0.8	1.2	2.1

Table 9.5 Fitted odds ratio—QS(1)

		Leadership								
		1	2	3	4	5	6	7	8	9
Literacy	1	7.1	**0.4**	1.0	0.6	0.9	0.9	0.4	3.9	1.1
	2	**0.4**	1.6	**0.6**	0.6	6.0	0.6	4.9	0.2	0.4
	3	1.0	**0.6**	2.0	**8.1**	0.2	1.2	0.7	0.8	1.9
	4	1.0	0.6	**8.1**	0.6	**0.3**	2.2	0.7	2.4	0.3
	5	0.5	10.0	0.2	**0.3**	5.1	**0.8**	1.0	0.7	2.1
	6	2.1	0.3	2.1	2.2	**0.8**	1.1	**0.7**	1.2	1.1
	7	0.2	12.5	0.4	1.2	1.0	**0.7**	1.7	**1.1**	0.8
	8	7.3	0.1	1.9	1.2	1.1	1.2	**1.1**	0.8	**1.2**
	9	0.6	0.8	1.1	0.8	1.1	1.8	0.8	**1.2**	2.1

Table 9.6 Fitted odds ratio—QS(2)

		Literacy								
		1	2	3	4	5	6	7	8	9
Literacy	1	7.18	**0.40**	**0.97**	0.56	1.15	0.63	0.87	1.96	1.20
	2	**0.40**	1.25	**0.80**	**0.75**	3.19	0.88	2.18	0.43	0.42
	3	**0.97**	**0.80**	1.64	**6.30**	**0.36**	1.20	0.74	0.74	1.73
	4	0.97	**0.75**	**6.30**	0.58	**0.25**	**2.22**	0.77	2.31	0.35
	5	0.56	5.51	**0.36**	**0.25**	5.52	**0.78**	**0.87**	0.70	2.10
	6	1.43	0.43	2.07	**2.22**	**0.78**	1.09	**0.73**	**1.26**	1.02
	7	0.48	4.96	0.36	1.33	**0.87**	**0.73**	1.65	**1.17**	**0.79**
	8	4.58	0.24	1.68	1.13	1.21	**1.26**	**1.17**	0.80	**1.25**
	9	0.58	0.99	0.95	0.80	1.02	1.77	**0.79**	**1.25**	2.12

Table 9.7 Fit statistics of quasi-symmetry and quasi-symmetry with degree n

Model	Df	G^2	p-value	AIC	BIC
QS	36	23.20	0.9510	−48.80	**−246.89**
QS(1)	29	1.74	1.0000	−56.26	−215.84
QS(2)	30	1.87	1.0000	**−58.13**	−223.21
QS(3)	31	7.04	1.0000	−54.96	−225.54
QS(4)	32	7.31	1.0000	−56.69	−232.78
QS(5)	33	13.31	0.9991	−52.69	−234.28
QS(6)	34	14.28	0.9988	−53.72	−240.81
QS(7)	35	23.20	0.9510	−48.80	−246.89

Table 9.7 lists the fit indices of QS and the series of QS(n) model. All the series of QS(n) fit well (All G^2 show non-significant results). QS has the best fit in BIC, whereas QS(2) shows the best fit in AIC.

The following R syntax put all the fit statistics together to produce the Table 9.7.

```
QSFit  <- model.summary(QS)
QS1Fit <- model.summary(QS1)
QS2Fit <- model.summary(QS2)
QS3Fit <- model.summary(QS3)
QS4Fit <- model.summary(QS4)
QS5Fit <- model.summary(QS5)
QS6Fit <- model.summary(QS6)
QS7Fit <- model.summary(QS7)
QSFitAll <- rbind(QSFit[1,1:5],
                  QS1Fit[1,1:5],
                  QS2Fit[1,1:5],
                  QS3Fit[1,1:5],
                  QS4Fit[1,1:5],
                  QS5Fit[1,1:5],
                  QS6Fit[1,1:5],
                  QS7Fit[1,1:5])
rownames(QSFitAll)
<- c("QS","QS(1)","QS(2)","QS(3)","QS(4)","QS(6)","QS(6)","QS(7)")
QSFitAll
```

```
   QSFitAll
        Df    G2      p    AIC     BIC
S       36 23.20 0.9510 -48.80 -246.89
S(1)    29  1.74 1.0000 -56.26 -215.84
S(2)    30  1.87 1.0000 -58.13 -223.21
S(3)    31  7.04 1.0000 -54.96 -225.54
S(4)    32  7.31 1.0000 -56.69 -232.78
S(6)    33 13.31 0.9991 -52.69 -234.28
S(6)    34 14.28 0.9988 -53.72 -240.81
S(7)    36 23.20 0.9510 -48.80 -246.89
```

Exercise

9.1 The following two tables (9.8 and 9.9) tabulate two scales A and B. The first table gives the cross-tabulation of A and B while the second table gives the cross-tabulation of A and the reversed scale of B. Generate complete symmetry for the first table and reverse complete symmetry model for the second table and comment.

9.2 The following table (9.10) is the cross-tabulation of literacy and numeracy skills. These two scales range from 1 to 10 being 1 representing low in literacy or numeracy skills, and 10 representing high in literacy or numeracy skills. Generate QS(n) models and comment.

Table 9.8 Scale A × Scale B

Scale A	Scale B			
	1	2	3	4
1	63	69	41	99
2	67	52	70	50
3	45	72	65	52
4	97	52	50	78

Table 9.9 Scale A × Scale B reversed

Scale A	Scale B			
	4	3	2	1
1	99	41	69	63
2	50	70	52	67
3	52	65	72	45
4	78	50	52	97

Table 9.10 Literacy × Numeracy skills

		Numeracy									
		1	2	3	4	5	6	7	8	9	10
Literacy	1	23	13	3	8	4	2	4	0	1	0
	2	11	4	0	5	3	1	2	1	0	0
	3	15	3	3	5	7	3	7	2	2	4
	4	28	11	12	28	22	14	18	11	6	4
	5	12	11	8	32	13	13	15	9	8	9
	6	12	15	15	32	29	23	19	13	10	26
	7	19	16	13	53	39	33	116	35	24	49
	8	10	5	9	14	11	16	28	28	27	32
	9	8	1	5	15	20	24	37	35	32	33
	10	13	4	10	19	24	27	65	50	45	134

9.3   The following table (9.11) is the cross-tabulation of literacy and leadership skills. These two scales range from 1 to 10 being 1 representing low in literacy or leadership skills, and 10 representing high in literacy or leadership skills. Generate QS(n) models and comment.

9.4   The following table (9.12) is the cross-tabulation of literacy and problem-solving skills. These two scales range from 1 to 10 being 1 representing low in literacy or problem solving skills, and 10 representing high in literacy or problem-solving skills. Generate QS(n) models and comment.

Table 9.11 Literacy × Leadership skills

		Leadership									
		1	2	3	4	5	6	7	8	9	10
Literacy	1	31	2	3	5	0	5	6	2	1	3
	2	8	4	3	4	3	1	1	1	0	2
	3	21	4	5	3	1	5	2	5	1	4
	4	30	6	6	8	11	17	26	16	15	19
	5	26	1	1	17	13	7	26	10	15	14
	6	26	3	9	31	5	14	35	14	26	31
	7	48	6	10	36	9	27	76	26	44	115
	8	20	0	6	14	6	13	22	13	29	57
	9	20	4	7	9	7	8	29	16	30	80
	10	42	7	5	23	9	16	37	16	33	203

Table 9.12 Literacy × Leadership skills

		Problem solving									
		1	2	3	4	5	6	7	8	9	10
Literacy	1	6	3	4	6	2	2	8	6	4	17
	2	2	0	0	3	4	2	1	2	6	7
	3	0	0	0	3	3	4	12	8	8	13
	4	3	2	2	7	8	11	28	18	22	53
	5	2	1	0	6	16	11	30	15	22	27
	6	1	2	1	6	11	9	44	29	33	58
	7	0	0	4	5	6	22	84	46	59	171
	8	0	0	0	0	4	7	19	26	34	90
	9	1	0	0	1	4	6	20	22	42	114
	10	1	2	1	1	7	4	25	34	33	283

Reference

Tan, T. K., & Sheng, Y. Z. (2015). Extending the quasi-symmetry model: Quasi-symmetry with n degree. Poster presented at the useR! Conference 2015

Chapter 10
Summary

10.1 Symbolic Table Summary

This section summarizes the symbolic tables discussed for the various chapters. Tables 10.1, 10.2, 10.3, and 10.4 tabulate the summary for asymmetry models, point symmetry models, non-independence models, and asymmetry + non-independence models respectively.

10.2 R Syntax Summary

The following four tables (Tables 10.5, 10.6, 10.7, and 10.8) summarize the syntax for symmetry/asymmetry, point symmetry, non-independence, asymmetry + non-independence models, respectively, for the doubly classified models discussed in the various chapters. The matrix form of the factor and regression variables are listed in Table 10.9.

10.3 Hierarchical Tree

Many statistical models can be classified as special cases of a more general statistical model. Doubly classified models are not exception to this general rule in classification. When these models are linked, their associations become clearer. There are sections in the various chapters illustrated the relationships of a few models to show their associations. For instance, Sect. 4.8 shows the association of conditional symmetry, linear diagonals parameters symmetry model, and 2-ratios parameters symmetry model, and Sect. 5.2.1 shows the relationship between complete point symmetry and inclined point symmetry model. However, they are not complete. This section puts all the models in a tree structure to show the

© Springer Nature Singapore Pte Ltd. 2017
T.K. Tan, *Doubly Classified Model with R*,
https://doi.org/10.1007/978-981-10-6995-6_10

Table 10.1 Summary of asymmetry models

Model	Description	Table
Complete symmetry	Symmetry in probability of off-diagonal cells	
Reverse complete symmetry	Symmetry in probability of reverse off-diagonal cells that run from top right to bottom left	
Conditional symmetry	Individual cell from the lower diagonal is a constant to that of the upper diagonal The total sum of the lower diagonal cells is also a constant to the total sum of the upper diagonal cells	
Odds symmetry I	The odds is a constant term for those cells enclosed with the same shape	
Odds symmetry II	The odds is a constant term for those cells enclosed with the same shape	
Parallel diagonal	Off-diagonal cells of the same position away from the diagonal are with the same probability	
Diagonal parameters symmetry	Cells with the same position away from the diagonal have the same odds	
Linear diagonal parameters symmetry	Cells with the same position away from the diagonal have the same odds, and the odds have a linear pattern	
Quasi symmetry	The odds ratios are in symmetry	
Quasi diagonal parameters symmetry	Difference in local odds ratios are the same for symmetrical cells enclosed by the same shape	

(continued)

Table 10.1 (continued)

2-Ratios parameter symmetry	The odds ratios are symmetrical for those cells that are two positions away from the diagonal (cells in red triangles), and those cells one position away are with a constant ratio γ The odds of those cells in the same position away from the diagonal are with the same odds and in linear form	<table><tr><td></td><td>1</td><td>2</td><td>3</td><td>4</td><td>5</td></tr><tr><td>1</td><td>γ</td><td>θ_{12}</td><td>θ_{13}</td><td>θ_{14}</td><td>θ_{15}</td></tr><tr><td>2</td><td>θ_{21}</td><td>γ</td><td>θ_{23}</td><td>θ_{24}</td><td>θ_{25}</td></tr><tr><td>3</td><td>θ_{31}</td><td>θ_{32}</td><td>γ</td><td>θ_{34}</td><td>θ_{35}</td></tr><tr><td>4</td><td>θ_{41}</td><td>θ_{42}</td><td>θ_{43}</td><td>γ</td><td>θ_{45}</td></tr><tr><td>5</td><td>θ_{51}</td><td>θ_{52}</td><td>θ_{53}</td><td>θ_{54}</td><td></td></tr></table> <table><tr><td></td><td>1</td><td>2</td><td>3</td><td>4</td><td>5</td><td>6</td></tr><tr><td>1</td><td>δ</td><td>π_{12}</td><td>π_{13}</td><td>π_{14}</td><td>π_{15}</td><td>π_{16}</td></tr><tr><td>2</td><td>π_{21}</td><td>δ</td><td>π_{23}</td><td>π_{24}</td><td>π_{25}</td><td>π_{26}</td></tr><tr><td>3</td><td>π_{31}</td><td>π_{32}</td><td>δ</td><td>π_{34}</td><td>π_{35}</td><td>π_{36}</td></tr><tr><td>4</td><td>π_{41}</td><td>π_{42}</td><td>π_{43}</td><td></td><td>π_{45}</td><td>π_{46}</td></tr><tr><td>5</td><td>π_{51}</td><td>π_{52}</td><td>π_{53}</td><td>π_{54}</td><td></td><td>π_{56}</td></tr><tr><td>6</td><td>π_{61}</td><td>π_{62}</td><td>π_{63}</td><td>π_{64}</td><td>π_{65}</td><td></td></tr></table>
Quasi conditional symmetry	The odds ratios are in symmetrical (cells in red triangles) except those cells that are one position away from the diagonal with a constant term γ Those cells with triangles over cells with circles are in a constant term γ	<table><tr><td></td><td>1</td><td>2</td><td>3</td><td>4</td><td>5</td></tr><tr><td>1</td><td>γ</td><td>θ_{12}</td><td>θ_{13}</td><td>θ_{14}</td><td>θ_{15}</td></tr><tr><td>2</td><td>θ_{21}</td><td>γ</td><td>θ_{23}</td><td>θ_{24}</td><td>θ_{25}</td></tr><tr><td>3</td><td>θ_{31}</td><td>θ_{32}</td><td>γ</td><td>θ_{34}</td><td>θ_{35}</td></tr><tr><td>4</td><td>θ_{41}</td><td>θ_{42}</td><td>θ_{43}</td><td>γ</td><td>θ_{45}</td></tr><tr><td>5</td><td>θ_{51}</td><td>θ_{52}</td><td>θ_{53}</td><td>θ_{54}</td><td></td></tr></table> △ = ▽ <table><tr><td></td><td>1</td><td>2</td><td>3</td><td>4</td><td>5</td><td>6</td></tr><tr><td>1</td><td></td><td>π_{12}</td><td>π_{13}</td><td>π_{14}</td><td>π_{15}</td><td>π_{16}</td></tr><tr><td>2</td><td>π_{21}</td><td></td><td>π_{23}</td><td>π_{24}</td><td>π_{25}</td><td>π_{26}</td></tr><tr><td>3</td><td>π_{31}</td><td>π_{32}</td><td></td><td>π_{34}</td><td>π_{35}</td><td>π_{36}</td></tr><tr><td>4</td><td>π_{41}</td><td>π_{42}</td><td>π_{43}</td><td></td><td>π_{45}</td><td>π_{46}</td></tr><tr><td>5</td><td>π_{51}</td><td>π_{52}</td><td>π_{53}</td><td>π_{54}</td><td></td><td>π_{56}</td></tr><tr><td>6</td><td>π_{61}</td><td>π_{62}</td><td>π_{63}</td><td>π_{64}</td><td>π_{65}</td><td></td></tr></table>
Quasi odds symmetry	The odds ratios are in symmetrical except those cells one position away from the diagonal. The constant term γ_j describes the relationship of odds ratios of these cells Those cells with triangles over cells with circles are in a constant term γ_j	<table><tr><td></td><td>1</td><td>2</td><td>3</td><td>4</td><td>5</td></tr><tr><td>1</td><td>γ</td><td>θ_{12}</td><td>θ_{13}</td><td>θ_{14}</td><td>θ_{15}</td></tr><tr><td>2</td><td>θ_{21}</td><td>γ</td><td>θ_{23}</td><td>θ_{24}</td><td>θ_{25}</td></tr><tr><td>3</td><td>θ_{31}</td><td>θ_{32}</td><td>γ</td><td>θ_{34}</td><td>θ_{35}</td></tr><tr><td>4</td><td>θ_{41}</td><td>θ_{42}</td><td>θ_{43}</td><td>γ</td><td>θ_{45}</td></tr><tr><td>5</td><td>θ_{51}</td><td>θ_{52}</td><td>θ_{53}</td><td>θ_{54}</td><td></td></tr></table> <table><tr><td></td><td>1</td><td>2</td><td>3</td><td>4</td><td>5</td><td>6</td></tr><tr><td>1</td><td></td><td>π_{12}</td><td>π_{13}</td><td>π_{14}</td><td>π_{15}</td><td>π_{16}</td></tr><tr><td>2</td><td>π_{21}</td><td></td><td>π_{23}</td><td>π_{24}</td><td>π_{25}</td><td>π_{26}</td></tr><tr><td>3</td><td>π_{31}</td><td>π_{32}</td><td>γ</td><td>π_{34}</td><td>π_{35}</td><td>π_{36}</td></tr><tr><td>4</td><td>π_{41}</td><td>π_{42}</td><td>π_{43}</td><td></td><td>π_{45}</td><td>π_{46}</td></tr><tr><td>5</td><td>π_{51}</td><td>π_{52}</td><td>π_{53}</td><td>π_{54}</td><td></td><td>π_{56}</td></tr><tr><td>6</td><td>π_{61}</td><td>π_{62}</td><td>π_{63}</td><td>π_{64}</td><td>π_{65}</td><td></td></tr></table>
Quasi symmetry with degree 1	Those cells one position off-diagonal are with the same odds ratios	<table><tr><td></td><td>1</td><td>2</td><td>3</td><td>4</td><td>5</td></tr><tr><td>1</td><td></td><td>θ_{12}</td><td></td><td></td><td></td></tr><tr><td>2</td><td>θ_{21}</td><td></td><td>θ_{23}</td><td></td><td></td></tr><tr><td>3</td><td></td><td>θ_{32}</td><td></td><td>θ_{34}</td><td></td></tr><tr><td>4</td><td></td><td></td><td>θ_{43}</td><td></td><td>θ_{45}</td></tr><tr><td>5</td><td></td><td></td><td></td><td>θ_{54}</td><td></td></tr></table>
Quasi symmetry with degree 2	Those cells one and two positions off-diagonal are with the same odds ratios	<table><tr><td></td><td>1</td><td>2</td><td>3</td><td>4</td><td>5</td></tr><tr><td>1</td><td></td><td>θ_{12}</td><td>θ_{13}</td><td></td><td></td></tr><tr><td>2</td><td>θ_{21}</td><td></td><td>θ_{23}</td><td>θ_{24}</td><td></td></tr><tr><td>3</td><td>θ_{31}</td><td>θ_{32}</td><td></td><td>θ_{34}</td><td>θ_{35}</td></tr><tr><td>4</td><td></td><td>θ_{42}</td><td>θ_{43}</td><td></td><td>θ_{45}</td></tr><tr><td>5</td><td></td><td></td><td>θ_{53}</td><td>θ_{54}</td><td></td></tr></table>
Quasi symmetry with degree 3	Those cells one, two, and three positions off-diagonal are with the same odds ratios	<table><tr><td></td><td>1</td><td>2</td><td>3</td><td>4</td><td>5</td></tr><tr><td>1</td><td></td><td>θ_{12}</td><td>θ_{13}</td><td>θ_{14}</td><td></td></tr><tr><td>2</td><td>θ_{21}</td><td></td><td>θ_{23}</td><td>θ_{24}</td><td>θ_{25}</td></tr><tr><td>3</td><td>θ_{31}</td><td>θ_{32}</td><td></td><td>θ_{34}</td><td>θ_{35}</td></tr><tr><td>4</td><td>θ_{41}</td><td>θ_{42}</td><td>θ_{43}</td><td></td><td>θ_{45}</td></tr><tr><td>5</td><td></td><td>θ_{52}</td><td>θ_{53}</td><td>θ_{54}</td><td></td></tr></table>
Marginal symmetry	Symmetry in marginal total	<table><tr><td></td><td>1</td><td>2</td><td>3</td><td>4</td><td>5</td><td>Total</td></tr><tr><td>1</td><td></td><td></td><td></td><td></td><td></td><td>$\pi_{1\cdot}$</td></tr><tr><td>2</td><td></td><td></td><td></td><td></td><td></td><td>$\pi_{2\cdot}$</td></tr><tr><td>3</td><td></td><td></td><td></td><td></td><td></td><td>$\pi_{3\cdot}$</td></tr><tr><td>4</td><td></td><td></td><td></td><td></td><td></td><td>$\pi_{4\cdot}$</td></tr><tr><td>5</td><td></td><td></td><td></td><td></td><td></td><td>$\pi_{5\cdot}$</td></tr><tr><td>Total</td><td>$\pi_{\cdot1}$</td><td>$\pi_{\cdot2}$</td><td>$\pi_{\cdot3}$</td><td>$\pi_{\cdot4}$</td><td>$\pi_{\cdot5}$</td><td></td></tr></table>

Table 10.2 Summary of point symmetry models

Model	Description	Table
Complete point symmetry	Symmetry in probabilities from the center of table. Odds ratios table also displays point symmetry to the center of the table	4×4 table of π_{ij} with point symmetry about Center; accompanying 3×3 table of θ_{ij}
Inclined point symmetry	Symmetry in probabilities with a constant ratio r from the center of the table. Diagonal cells symmetry with a constant ratio 1	4×4 table of π_{ij} with constant ratio r
Quasi point symmetry	Symmetry in odds ratios from the center of local odds ratio table	3×3 table of θ_{ij}
Quasi inclined point symmetry	Point symmetrical in odds ratios for those cells in circles. Those cells one position away from the diagonal (square cells) are with a constant φ	4×4 table of θ_{ij} with constant φ
Proportional point symmetry	Point symmetry in probabilities with a constant term α but not diagonal cells	4×4 table of π_{ij} with constant α
Local point symmetry	Symmetry in probabilities except the diagonal cells	4×4 table of π_{ij}; Not in Symmetry
Reverse local point symmetry	Symmetry in probabilities except the diagonal cells that run from top right to bottom left	4×4 table of π_{ij}; Not in Symmetry
Reverse proportional Point symmetry	Reverse point symmetry in probabilities with a constant term β but not reverse diagonal cells	4×4 table of π_{ij} with constant β
Reverse inclined point symmetry	The diagonal cells exhibit reverse symmetrical. The rest of the cells are symmetrical to the center with a constant γ	4×4 table of π_{ij} with constant γ
Quasi reverse inclined point symmetry	Those cells that are one position away are with a constant ϕ. The rest are in point symmetrical to the center	4×4 table of θ_{ij} with constant ϕ

(continued)

Table 10.2 (continued)

			1	2	3	4
Reverse conditional symmetry	The symmetrical pattern is shown in the square table on the right with a constant term η. The reverse diagonal cells are not in symmetrical	1	π_{11}	π_{12}	π_{13}	π_{14}
		2	π_{21}	π_{22}	π_{23}	π_{24}
		3	π_{31}	$\eta\pi_{32}$	π_{33}	π_{34}
		4	π_{41}	π_{42}	π_{43}	π_{44}

Table 10.3 Summary of non-independence models

Model	Description
Independence	Odds ratios are all ones
Fixed distance	A constant odds ratio for the diagonal and the rest of the odds ratios are 1s
Variable distance	Different odds ratios at the diagonal and the rest of the odds ratios are 1s
Uniform royalty	A constant odds ratio at the diagonal and another constant odds ratio for cells one position off the diagonal and the rest of the odds ratios are 1s
Quasi independence	Different odds ratios at the diagonal and symmetrical odds ratios for those cells one diagonal away from the diagonal and the rest of the odds ratios are 1s
Triangle parameter	A constant odds ratio at the diagonal. Odds ratios are the same for those cells one position off the upper diagonal as well as lower diagonal and the rest of the odds ratios are 1s
Uniform association	A constant odds ratio value
Diagonal D	The odds ratios of the diagonal cells are the same. For the off-diagonal cells, each cell has its own odds ratio
Diagonal absolute	Odds ratios are symmetrical and parallel to the diagonal, including the diagonal cells

Table representations for each model:

Independence

θ	1	2	3	4
1	1	1	1	1
2	1	1	1	1
3	1	1	1	1
4	1	1	1	1

Fixed distance

	1	2	3	4
1	θ	1	1	1
2	1	θ	1	1
3	1	1	θ	1
4	1	1	1	θ

Variable distance

	1	2	3	4
1	θ_{11}	1	1	1
2	1	θ_{22}	1	1
3	1	1	θ_{33}	1
4	1	1	1	θ_{44}

Uniform royalty

	1	2	3	4
1	θ_1	θ_2	1	1
2	θ_2	θ_1	θ_2	1
3	1	θ_2	θ_1	θ_2
4	1	1	θ_2	θ_1

Quasi independence

	1	2	3	4
1	θ_{11}	θ_2	1	1
2	θ_2	θ_{22}	θ_3	1
3	1	θ_3	θ_{33}	θ_4
4	1	1	θ_4	θ_{44}

Triangle parameter

	1	2	3	4
1	θ_1	θ_2	1	1
2	θ_3	θ_1	θ_2	1
3	1	θ_3	θ_1	θ_2
4	1	1	θ_3	θ_1

Uniform association

	1	2	3	4
1	θ	θ	θ	θ
2	θ	θ	θ	θ
3	θ	θ	θ	θ
4	θ	θ	θ	θ

Diagonal D

	1	2	3	4
1	θ_{11}	θ_{12}	θ_{13}	θ_{14}
2	θ_{21}	θ_{11}	θ_{23}	θ_{24}
3	θ_{31}	θ_{32}	θ_{11}	θ_{34}
4	θ_{41}	θ_{42}	θ_{43}	θ_{11}

Diagonal absolute

	1	2	3	4
1	θ_1	θ_2	θ_3	θ_4
2	θ_2	θ_1	θ_2	θ_3
3	θ_3	θ_2	θ_1	θ_2
4	θ_4	θ_3	θ_2	θ_1

(continued)

Table 10.3 (continued)

Uniform fixed distance association	Odds ratio at the diagonal has a common value while the off-diagonal another common odds ratio	

	1	2	3	4
1	θ_1	θ_2	θ_2	θ_2
2	θ_2	θ_1	θ_2	θ_2
3	θ_2	θ_2	θ_1	θ_2
4	θ_2	θ_2	θ_2	θ_1

Uniform variable distance association	Odds ratios at the diagonal vary while a common odds ratio for all the off-diagonal cells	

	1	2	3	4
1	θ_{11}	θ_2	θ_2	θ_2
2	θ_2	θ_{22}	θ_2	θ_2
3	θ_2	θ_2	θ_{33}	θ_2
4	θ_2	θ_2	θ_2	θ_{44}

Table 10.4 Summary of asymmetry + non-independence models

Model	Description	Table
Non-symmetry + independence	Odds ratios are ones	
Non-symmetry + independence triangle	A constant odds ratio at the diagonal. Odds ratios are the same for one cell position off the upper diagonal as well as lower diagonal and the rest of the odds ratios are 1s	
Non-symmetry + independence diagonals	Odds ratios are all in parallel to the diagonal and each parallel diagonal has a common odds ratio	
Non-symmetry + independence diagonals absolute	A constant odds ratio at the diagonal. Odds ratios with the same position away from diagonal have the same value	
Non-symmetry + independence diagonals absolute triangle	A constant odds ratio at the diagonal. Upper diagonal with one position from the diagonal has the same odds ratios. Lower diagonal with position from the diagonal has the same odds ratios. Those cells at two positions away from the diagonal have symmetrical odds ratios	

Non-symmetry + independence:

θ	1	2	3	4
1	1	1	1	1
2	1	1	1	1
3	1	1	1	1
4	1	1	1	1

Non-symmetry + independence triangle:

	1	2	3	4
1	θ_1	θ_2	1	1
2	θ_3	θ_1	θ_2	1
3	1	θ_3	θ_1	θ_2
4	1	1	θ_3	θ_1

Non-symmetry + independence diagonals:

	1	2	3	4
1	θ_1	θ_2	θ_3	θ_4
2	θ_5	θ_1	θ_2	θ_3
3	θ_6	θ_5	θ_1	θ_2
4	θ_7	θ_6	θ_5	θ_1

Non-symmetry + independence diagonals absolute:

	1	2	3	4
1	θ_1	θ_2	θ_3	θ_4
2	θ_2	θ_1	θ_2	θ_3
3	θ_3	θ_2	θ_1	θ_2
4	θ_4	θ_3	θ_2	θ_1

Non-symmetry + independence diagonals absolute triangle:

	1	2	3	4
1	θ_1	θ_2	θ_4	θ_5
2	θ_3	θ_1	θ_2	θ_4
3	θ_4	θ_3	θ_1	θ_2
4	θ_5	θ_4	θ_3	θ_1

Table 10.5 Syntax—symmetry/asymmetry models

Symmetry/Asymmetry model	Regression/Factor variable
Complete symmetry (S)	S
Reversed complete symmetry (RS)	RS
Conditional symmetry (CS)	S + CS
Odds symmetry I (OS1)	S + OS1
Odds symmetry II (OS2)	S + OS2
Parallel diagonal symmetry (PDS)	S + PD
Diagonal parameters symmetry (DPS)	S + D
Linear diagonal parameters symmetry (LDPS)	S + F
Quasi-symmetry (QS)	R + C + S
Quasi diagonal parameters symmetry (QDPS)	R + C + S + D
2-Ratios parameter symmetry (2RPS)	S + CS + F
Quasi conditional symmetry (QCS)	R + C + S + CS
Quasi odds symmetry (QOS)	R + C + S + OS1 or R + C + S + OS2
Quasi symmetry with degree 1 (QS1)	R + C + S + QS1
Quasi symmetry with degree 2 (QS2)	R + C + S + QS2
Quasi symmetry with degree 3 (QS3)	R + C + S + QS3
…	
Quasi symmetry with degree n (QSn)	R + C + S + QSn

Table 10.6 Syntax—point symmetry models

Point symmetry model	Regression/Factor variable
Complete point symmetry model	P
Inclined point symmetry model	P + Psi
Quasi point symmetry model	P + R + C
Quasi inclined point symmetry model	R + C + P + Psi
Proportional point symmetry model	PP + Psi
Local point symmetry model	PP
Reverse local point symmetry model	RL
Reverse proportional point symmetry model	RP + Delta
Reverse inclined point symmetry model	RI + Delta
Quasi reverse inclined point symmetry model	R + C + RI + Delta
Reverse conditional symmetry model	RC + Delta
Quasi reverse conditional symmetry model	R + C + RC + Delta

Table 10.7 Syntax—non-independence models

Non-independence models	Regression/Factor variable
Independence model	R + C
Principal diagonal models	
Fixed distance model	R + C + F
Variable distance model	R + C + V1 + V2 + V3 + ...
Diagonal band models	
Uniform loyalty model	R + C + L
Quasi independence model	R + C + Q
Triangle parameters model	R + C + T
Full diagonal model	
Diagonal D model /Asymmetric minor diagonal model	R + C + D
Diagonal absolute model	R + C + DA
Uniform association model	R + C + U
Uniform fixed distance association model	R + C + U + F
Uniform variable distance association model	R + C + U + V1 + V2 + V3 + ...

Table 10.8 Asymmetry + non-independence models

Asymmetry + non-independence models	Regression/Factor variable
Null model	H1 + H2 + H3 + H4 + ...
Asymmetry + non-independence triangle	H1 + H2 + H3 + H4 + ... +T
Asymmetry + non-independence diagonals	H1 + H2 + H3 + H4 + ... +D
Asymmetry + non-independence diagonals absolute	H1 + H2 + H3 + H4 + ... +DA
Asymmetry + non-independence diagonals absolute triangle	H1 + H2 + H3 + H4 + ... + DA + T

relationship of doubly classified models. As the association looks like a tree, it is referred to by Lawal (2000, 2004) as hierarchical tree to show the hierarchy of doubly classified models presented in a tree-like structure. Figures 10.1 and 10.2 show the hierarchical trees of asymmetry and point symmetry models, respectively (Tables 10.10, 10.11 and 10.12).

10.3.1 Hierarchical Tree—Asymmetry Models

The relationships of the asymmetry models could be further summarized as hierarchical tree, a tree structure that describes the nested association of models (e.g., Lawal 2000, 2004). Figure 10.2 gives the hierarchical tree of asymmetry models, adapted from Lawal (2004).

On the middle left-hand side of Fig. 10.2 is the simplest model, the complete symmetry model (*S*), that exhibits the total symmetry in probabilities of cell with the diagonal as the reference. The rest of the models exhibit some kinds of variation

Table 10.9 Asymmetry model factors and regression variables

$$
S = \begin{pmatrix}
1 & 2 & 3 & 4 & 5 & 6 \\
2 & 7 & 8 & 9 & 10 & 11 \\
3 & 8 & 12 & 13 & 14 & 15 \\
4 & 9 & 13 & 16 & 17 & 18 \\
5 & 10 & 14 & 17 & 19 & 20 \\
6 & 11 & 15 & 18 & 20 & 21
\end{pmatrix}
\qquad
RS = \begin{bmatrix}
6 & 5 & 4 & 3 & 2 & 1 \\
11 & 10 & 9 & 8 & 7 & 2 \\
15 & 14 & 13 & 12 & 8 & 3 \\
18 & 17 & 16 & 13 & 9 & 4 \\
20 & 19 & 17 & 14 & 10 & 5 \\
21 & 20 & 18 & 15 & 11 & 6
\end{bmatrix}
\qquad
D = \begin{pmatrix}
11 & 1 & 1 & 2 & 3 & 4 & 5 \\
6 & 11 & 6 & 1 & 2 & 3 & 4 \\
7 & 6 & 11 & 1 & 2 & 3 \\
8 & 7 & 6 & 8 & 7 & 6 & 2 \\
9 & 8 & 7 & 9 & 8 & 7 & 6 \\
10 & 9 & 8 & 7 & 6 & 11
\end{pmatrix}
$$

$$
OS1 = \begin{pmatrix}
11 & 1 & 1 & 1 & 1 & 6 \\
6 & 11 & 2 & 2 & 2 & 11 \\
6 & 7 & 11 & 3 & 3 & 15 \\
6 & 7 & 8 & 11 & 4 & 18 \\
6 & 7 & 8 & 9 & 11 & 20 \\
6 & 7 & 8 & 9 & 10 & 21
\end{pmatrix}
\qquad
OS2 = \begin{pmatrix}
11 & 5 & 4 & 3 & 2 & 1 \\
10 & 11 & 4 & 3 & 2 & 1 \\
9 & 9 & 11 & 3 & 2 & 1 \\
8 & 8 & 8 & 11 & 2 & 1 \\
7 & 7 & 7 & 7 & 11 & 1 \\
6 & 6 & 6 & 6 & 6 & 11
\end{pmatrix}
\qquad
CS = \begin{pmatrix}
1 & 1 & 1 & 1 & 1 & 1 \\
2 & 1 & 1 & 1 & 1 & 1 \\
2 & 2 & 1 & 1 & 1 & 1 \\
2 & 2 & 2 & 1 & 1 & 1 \\
2 & 2 & 2 & 2 & 1 & 1
\end{pmatrix}
$$

$$
F = \begin{bmatrix}
1 & 2 & 3 & 4 & 5 & 6 \\
1 & 2 & 3 & 4 & 5 & -1 \\
1 & 2 & 3 & 4 & -1 & -1 \\
1 & 2 & 3 & -1 & -1 & -1 \\
1 & 2 & -1 & -1 & -1 & -1 \\
1 & -1 & -1 & -1 & -1 & -1
\end{bmatrix}
\qquad
R = \begin{pmatrix}
1 & 2 & 3 & 4 & 5 & 6 \\
1 & 2 & 3 & 4 & 5 & 6 \\
1 & 2 & 3 & 4 & 5 & 6 \\
1 & 2 & 3 & 4 & 5 & 6 \\
1 & 2 & 3 & 4 & 5 & 6 \\
1 & 2 & 3 & 4 & 5 & 6
\end{pmatrix}
\qquad
PD = \begin{pmatrix}
11 & 10 & 9 & 8 & 7 & 6 \\
7 & 7 & 8 & 9 & 10 & 11 \\
7 & 7 & 8 & 9 & 10 & 11 \\
8 & 8 & 8 & 9 & 10 & 11 \\
9 & 9 & 9 & 9 & 10 & 11 \\
10 & 10 & 10 & 10 & 10 & 11
\end{pmatrix}
\qquad
D = \begin{pmatrix}
0 & -1 & -2 & -3 & -4 & -5 \\
1 & 0 & -1 & -2 & -3 & -4 \\
2 & 1 & 0 & -1 & -2 & -3 \\
3 & 2 & 1 & 0 & -1 & -2 \\
4 & 3 & 2 & 1 & 0 & -1 \\
5 & 4 & 3 & 2 & 1 & 0
\end{pmatrix}
$$

$$
QS2 = \begin{pmatrix}
1 & 2 & 3 & 4 & 4 & 4 \\
2 & 1 & 2 & 3 & 4 & 4 \\
3 & 2 & 1 & 2 & 3 & 4 \\
4 & 3 & 2 & 1 & 2 & 3 \\
5 & 4 & 3 & 2 & 1 & 2 \\
6 & 5 & 4 & 3 & 2 & 1
\end{pmatrix}
\qquad
QS3 = \begin{pmatrix}
1 & 2 & 3 & 4 & 5 & 5 \\
2 & 1 & 2 & 3 & 4 & 5 \\
3 & 2 & 1 & 2 & 3 & 4 \\
4 & 3 & 2 & 1 & 2 & 3 \\
5 & 4 & 3 & 2 & 1 & 2 \\
6 & 5 & 4 & 3 & 2 & 1
\end{pmatrix}
\qquad
C = \begin{pmatrix}
1 & 1 & 1 & 1 & 1 & 1 \\
2 & 2 & 2 & 2 & 2 & 2 \\
3 & 3 & 3 & 3 & 3 & 3 \\
4 & 4 & 4 & 4 & 4 & 4 \\
5 & 5 & 5 & 5 & 5 & 5 \\
6 & 6 & 6 & 6 & 6 & 6
\end{pmatrix}
\qquad
QS1 = \begin{pmatrix}
1 & 2 & 3 & 3 & 3 & 3 \\
2 & 1 & 2 & 3 & 3 & 3 \\
3 & 2 & 1 & 2 & 3 & 3 \\
4 & 3 & 2 & 1 & 2 & 3 \\
5 & 4 & 3 & 2 & 1 & 2 \\
6 & 5 & 4 & 3 & 2 & 1
\end{pmatrix}
$$

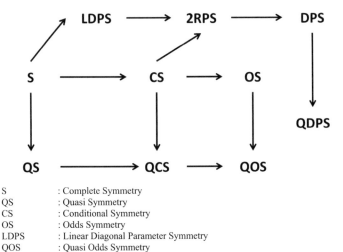

S	: Complete Symmetry	
QS	: Quasi Symmetry	
CS	: Conditional Symmetry	
OS	: Odds Symmetry	
LDPS	: Linear Diagonal Parameter Symmetry	
QOS	: Quasi Odds Symmetry	
QCS	: Quasi Conditional Symmetry	
DPS	: Diagonal Parameter Symmetry	
QDPS	: Quasi-Diagonal Parameter Symmetry	
2RPS	: 2 Ratio Parameter Symmetry	

Fig. 10.1 Hierarchical tree of asymmetry models

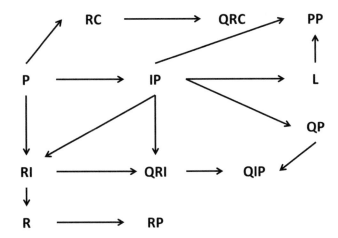

P	: Complete Point Symmetry
IP	: Inclined Point Symmetry
QIP	: Quasi Inclined Point Symmetry
QP	: Quasi Point Symmetry
PP	: Proportional Point Symmetry
L	: Local Point Symmetry
RI	: Reverse Inclined Point Symmetry
R	: Reverse Local Point Symmetry
RP	: Reverse Proportional Point Symmetry
RC	: Reverse Conditional Symmetry
QRI	: Quasi Reverse Inclined Point Symmetry
QRC	: Quasi Reverse Conditional Symmetry

Fig. 10.2 Hierarchical tree of point symmetry models

Table 10.10 Point symmetry factors and regression variables

$$
P = \begin{pmatrix}
1 & 4 & 5 & 6 & 7 & 8 \\
9 & 2 & 10 & 11 & 12 & 13 \\
14 & 15 & 3 & 16 & 17 & 18 \\
18 & 17 & 16 & 3 & 15 & 14 \\
13 & 12 & 11 & 10 & 2 & 9 \\
8 & 7 & 6 & 5 & 4 & 1
\end{pmatrix}
\qquad
\mathrm{Psi} = \begin{pmatrix}
3 & 1 & 1 & 1 & 1 & 1 \\
2 & 3 & 1 & 1 & 1 & 1 \\
2 & 2 & 3 & 1 & 1 & 1 \\
2 & 2 & 2 & 3 & 1 & 1 \\
2 & 2 & 2 & 2 & 3 & 1 \\
2 & 2 & 2 & 2 & 2 & 3
\end{pmatrix}
$$

$$
\mathrm{PP} = \begin{pmatrix}
1 & 7 & 8 & 9 & 10 & 11 \\
12 & 2 & 13 & 14 & 15 & 16 \\
17 & 18 & 3 & 19 & 20 & 21 \\
21 & 20 & 19 & 4 & 18 & 17 \\
16 & 15 & 14 & 13 & 5 & 12 \\
11 & 10 & 9 & 8 & 7 & 6
\end{pmatrix}
\qquad
\mathrm{RL} = \begin{pmatrix}
1 & 4 & 5 & 6 & 7 & 16 \\
8 & 2 & 9 & 10 & 17 & 11 \\
12 & 13 & 3 & 18 & 14 & 15 \\
15 & 14 & 19 & 3 & 13 & 12 \\
11 & 20 & 10 & 9 & 2 & 8 \\
21 & 7 & 6 & 5 & 4 & 1
\end{pmatrix}
$$

$$
\delta = \begin{pmatrix}
3 & 3 & 3 & 3 & 3 & 1 \\
3 & 3 & 3 & 3 & 1 & 2 \\
3 & 3 & 3 & 1 & 2 & 2 \\
3 & 3 & 1 & 2 & 2 & 2 \\
3 & 1 & 2 & 2 & 2 & 2 \\
1 & 2 & 2 & 2 & 2 & 2
\end{pmatrix}
\qquad
\mathrm{RI} = \begin{pmatrix}
1 & 4 & 5 & 6 & 7 & 16 \\
8 & 2 & 9 & 10 & 17 & 11 \\
12 & 13 & 3 & 18 & 14 & 15 \\
15 & 14 & 18 & 3 & 13 & 12 \\
11 & 17 & 10 & 9 & 2 & 8 \\
16 & 7 & 6 & 5 & 4 & 1
\end{pmatrix}
$$

$$
\mathrm{RC} = \begin{pmatrix}
1 & 4 & 5 & 6 & 7 & 16 \\
8 & 2 & 9 & 10 & 17 & 7 \\
11 & 12 & 3 & 18 & 10 & 6 \\
13 & 14 & 19 & 3 & 9 & 5 \\
15 & 20 & 14 & 12 & 2 & 8 \\
21 & 7 & 6 & 5 & 4 & 1
\end{pmatrix}
$$

Table 10.11 Non-independence factors and regression variables

$$V1 = \begin{pmatrix} 2 & 1 & 1 & 1 \\ 1 & 2 & 2 & 2 \\ 1 & 2 & 2 & 2 \\ 1 & 2 & 2 & 2 \\ 1 & 2 & 2 & 2 \\ 1 & 2 & 2 & 2 \\ 1 & 2 & 2 & 2 \\ 1 & 2 & 2 & 2 \end{pmatrix} \qquad
V2 = \begin{pmatrix} 3 & 3 & 1 & 1 & 1 & 1 \\ 3 & 3 & 1 & 1 & 1 & 1 \\ 1 & 1 & 3 & 3 & 3 & 3 \\ 1 & 1 & 3 & 3 & 3 & 3 \\ 1 & 1 & 3 & 3 & 3 & 3 \\ 1 & 1 & 3 & 3 & 3 & 3 \\ 1 & 1 & 3 & 3 & 3 & 3 \\ 1 & 1 & 3 & 3 & 3 & 2 \end{pmatrix}$$

$$V3 = \begin{pmatrix} 4 & 4 & 4 & 1 & 1 & 1 \\ 4 & 4 & 4 & 1 & 1 & 1 \\ 4 & 4 & 4 & 1 & 1 & 1 \\ 1 & 1 & 1 & 4 & 4 & 4 \\ 1 & 1 & 1 & 4 & 4 & 4 \\ 1 & 1 & 1 & 4 & 4 & 4 \end{pmatrix} \qquad
V4 = \begin{pmatrix} 5 & 5 & 5 & 5 & 1 & 1 \\ 5 & 5 & 5 & 5 & 1 & 1 \\ 5 & 5 & 5 & 5 & 1 & 1 \\ 5 & 5 & 5 & 5 & 1 & 1 \\ 1 & 1 & 1 & 1 & 5 & 5 \\ 1 & 1 & 1 & 1 & 5 & 5 \end{pmatrix}$$

$$V5 = \begin{pmatrix} 6 & 6 & 6 & 1 \\ 6 & 6 & 6 & 1 \\ 6 & 6 & 6 & 1 \\ 6 & 6 & 6 & 1 \\ 6 & 6 & 6 & 1 \\ 6 & 6 & 6 & 1 \\ 6 & 6 & 1 & 1 \\ 1 & 1 & 1 & 6 \end{pmatrix} \qquad
L = \begin{pmatrix} 2 & 1 & 1 & 1 & 1 & 1 \\ 1 & 2 & 1 & 1 & 1 & 1 \\ 1 & 1 & 2 & 1 & 1 & 1 \\ 1 & 1 & 1 & 2 & 1 & 1 \\ 1 & 1 & 1 & 1 & 2 & 1 \\ 1 & 1 & 1 & 1 & 1 & 2 \end{pmatrix}$$

$$Q = \begin{pmatrix} 2 & 1 & 1 & 1 & 1 & 1 \\ 1 & 3 & 1 & 1 & 1 & 1 \\ 1 & 1 & 4 & 1 & 1 & 1 \\ 1 & 1 & 1 & 5 & 1 & 1 \\ 1 & 1 & 1 & 6 & 6 & 1 \\ 1 & 1 & 1 & 1 & 1 & 7 \end{pmatrix} \qquad
T = \begin{pmatrix} 3 & 1 & 1 & 1 & 1 & 1 \\ 2 & 3 & 1 & 1 & 1 & 1 \\ 2 & 2 & 3 & 1 & 1 & 1 \\ 2 & 2 & 2 & 3 & 1 & 1 \\ 2 & 2 & 2 & 2 & 3 & 1 \\ 2 & 2 & 2 & 2 & 2 & 3 \end{pmatrix}$$

$$D = \begin{pmatrix} 1 & 7 & 8 & 9 & 10 & 11 \\ 2 & 1 & 7 & 8 & 9 & 10 \\ 3 & 2 & 1 & 7 & 8 & 9 \\ 4 & 3 & 2 & 1 & 7 & 8 \\ 5 & 4 & 3 & 2 & 1 & 7 \\ 6 & 5 & 4 & 3 & 2 & 1 \end{pmatrix} \qquad
DA = \begin{pmatrix} 1 & 2 & 3 & 4 & 5 & 6 \\ 2 & 1 & 2 & 3 & 4 & 5 \\ 3 & 2 & 1 & 2 & 3 & 4 \\ 4 & 3 & 2 & 1 & 2 & 3 \\ 5 & 4 & 3 & 2 & 1 & 2 \\ 6 & 5 & 4 & 3 & 2 & 1 \end{pmatrix}$$

$$U = \begin{pmatrix} 1 & 2 & 3 & 4 & 5 & 6 \\ 2 & 4 & 6 & 8 & 10 & 12 \\ 3 & 6 & 9 & 12 & 15 & 18 \\ 4 & 8 & 12 & 16 & 20 & 24 \\ 5 & 10 & 15 & 20 & 25 & 30 \\ 6 & 12 & 18 & 24 & 30 & 35 \end{pmatrix}$$

Table 10.12 Asymmetry + non-independence factors and regression variables

$$
f_1 = \begin{pmatrix}
1 & 0 & 0 & 0 & 0 & 0 \\
1 & 1 & 1 & 1 & 1 & 1 \\
1 & 1 & 1 & 1 & 1 & 1 \\
1 & 1 & 1 & 1 & 1 & 1 \\
1 & 1 & 1 & 1 & 1 & 1 \\
1 & 0 & 0 & 0 & 0 & 0
\end{pmatrix}
\qquad
f_2 = \begin{pmatrix}
0 & 1 & 0 & 0 & 0 & 0 \\
0 & 1 & 0 & 0 & 0 & 0 \\
0 & 1 & 0 & 0 & 0 & 0 \\
0 & 1 & 0 & 0 & 0 & 0 \\
0 & 1 & 0 & 0 & 0 & 0 \\
0 & 1 & 0 & 0 & 0 & 1
\end{pmatrix}
\qquad
f_3 = \begin{pmatrix}
0 & 0 & 1 & 0 & 0 & 0 \\
0 & 0 & 1 & 0 & 0 & 0 \\
0 & 0 & 1 & 0 & 0 & 0 \\
0 & 0 & 1 & 0 & 0 & 0 \\
0 & 0 & 1 & 0 & 0 & 0 \\
0 & 0 & 1 & 0 & 0 & 0
\end{pmatrix}
\qquad
s_1 = \begin{pmatrix}
0 & 0 & 0 & 0 & 0 & 1 \\
0 & 0 & 0 & 0 & 0 & 1 \\
0 & 0 & 0 & 0 & 0 & 1 \\
0 & 0 & 0 & 0 & 0 & 1 \\
0 & 0 & 0 & 0 & 0 & 1 \\
0 & 0 & 0 & 0 & 0 & 1
\end{pmatrix}
$$

$$
s_2 = \begin{pmatrix}
0 & 0 & 0 & 0 & 0 & 0 \\
0 & 0 & 0 & 0 & 0 & 0 \\
0 & 1 & 1 & 1 & 1 & 1 \\
0 & 1 & 1 & 1 & 1 & 1 \\
0 & 1 & 1 & 1 & 1 & 1 \\
0 & 0 & 0 & 0 & 0 & 0
\end{pmatrix}
\qquad
s_3 = \begin{pmatrix}
0 & 0 & 0 & 0 & 0 & 0 \\
0 & 0 & 0 & 0 & 0 & 0 \\
0 & 1 & 0 & 0 & 1 & 1 \\
0 & 0 & 0 & 0 & 0 & 0 \\
0 & 0 & 0 & 0 & 0 & 0 \\
0 & 0 & 0 & 0 & 0 & 0
\end{pmatrix}
\qquad
H_1 = \begin{pmatrix}
2 & 1 & 1 & 1 & 1 & 1 \\
-1 & 1 & 1 & 1 & 1 & 1 \\
-1 & 1 & 0 & 0 & 0 & 0 \\
-1 & 1 & 0 & 0 & 0 & 0 \\
-1 & 1 & 0 & 0 & 0 & 0 \\
-1 & 1 & 0 & 0 & 0 & 0
\end{pmatrix}
\qquad
H_2 = \begin{pmatrix}
1 & -1 & 0 & 0 & 1 & -1 \\
-1 & 0 & 1 & 1 & 1 & 0 \\
0 & -1 & 1 & 0 & 0 & 0 \\
2 & -1 & 1 & 0 & 0 & 0 \\
-1 & -1 & 0 & 1 & 0 & 0 \\
-1 & -1 & 0 & 0 & 0 & 0
\end{pmatrix}
$$

$$
H_3 = \begin{pmatrix}
0 & 0 & -1 & 0 & 0 & 0 \\
0 & -1 & 0 & 0 & 0 & 0 \\
0 & -1 & 2 & 1 & 0 & 0 \\
0 & -1 & 1 & 0 & 0 & 1 \\
0 & -1 & 1 & 0 & 0 & 0 \\
0 & 0 & 1 & 0 & 0 & 0
\end{pmatrix}
$$

from this base model. The conditional symmetry model (CS) put in a constant term γ to describe the association between the right off-diagonal cells to the left off-diagonal cells. The arrow that points from S to CS shows that CS is a general model and S a more restrictive model with $\gamma = 1$. The odds symmetry models (OS) further relaxes this assumption by varying the constant γ into numerous terms depending on the rows or columns of cell position. The rest of the models are described by the hierarchical tree below.

10.3.2   Hierarchical Tree—Point Symmetry Models

Similar to the hierarchical tree of the asymmetry model, the hierarchical tree of point symmetry models is shown in Fig. 8.2. The basic model is the point symmetry model (P). The hierarchical tree is adapted from Lawal (2000).

10.4   Nested Models—Chi-Square Difference Test

For models that are nested, Chi-Square difference test can be carried out to examine whether a simple or a more complex model is supported. The hierarchical tree provides the reference for carrying out Chi-Square difference test. For instance, QS is a general model of S, and CS is also a general model of S. We can carry out a Chi-Square difference test between QS and S, and CS and S separately. Another example is we can carry out a test of independence based on the difference between G^2 values for the independence model and the uniform association model as shown in the hypothesis stated below (Simonoff 2003). The degree of freedom is one.

$$H_0 : \theta = 0$$

$$G^2(I|U) = G^2(I) - G^2(U)$$

The following two examples illustrated how to carry out Chi-Square difference test using R.

10.4.1   Chi-Square Difference Test—Complete Symmetry and Conditional Symmetry Model

Complete symmetry model states that $H_S : \pi_{ij} = \pi_{ji}$, and conditional symmetry model states that $H_{CS} : \pi_{ij} = \gamma \pi_{ji}$. When $\gamma = 1$, conditional symmetry model reduces to complete symmetry model. As the two models are nested, Chi-Square difference test is carried out to test $H_0 : = \gamma = 1$ (Tables 10.13, 10.14, 10.15, and 10.16).

Table 10.13 Residence mobility pattern

Last residence	Current residence			
	Ang Mo Kio	Toa payoh	Yishun	Sembawang
Ang Mo Kio	150	66	24	33
Toa payoh	68	12	32	81
Yishun	20	32	72	20
Sembawang	38	82	12	42

Table 10.14 Fit statistics—comparing complete and conditional symmetry model

Model	G^2	df	P-value	AIC	BIC
Complete symmetry	2.77	6	0.8366	−9.23	−37.21
Conditional symmetry	2.74	5	0.7396	−7.26	−30.58

Table 10.15 Fit statistics—comparing complete and conditional symmetry model

Model	G^2	df	P-value	AIC	BIC
Complete point	15.42	8	0.0515	−0.58	−37.92
Inclined point	0.26	7	0.9999	−13.74	−46.41

Table 10.16 Doctor A and B assessment on patients

Doctor A	Doctor B			
	1	2	3	4
1	42	37	55	99
2	42	32	32	87
3	60	25	33	57
4	73	42	27	43

The R syntax below provides the syntax for Chi-Square difference test.

```
DataS <- c(150, 66, 24, 33,
            68, 12, 32, 81,
            20, 32, 72, 20,
            38, 82, 12, 42)
s<-c( 1, 2, 3, 4,
      2, 5, 6, 7,
      3, 6, 8, 9,
      4, 7, 9,10)
s<-as.factor(s)
```

```
CS <- c( 1, 1, 1, 1,
         2, 1, 1, 1,
         2, 2, 1, 1,
         2, 2, 2, 1)
CS <- as.factor(CS)
CompS <- glm(DataS~s,family=poisson)
CondS <- glm(DataS~s+CS,family=poisson)
# Chi-Square Difference Test
df        <- CompS$df.residual-CondS$df.residual
deviance <- CompS$deviance-CondS$deviance
p         <- pchisq(deviance,160;df, lower.tail=F)
ChiSqDiff <- cbind(round(deviance,4), df, round(p,4))
colnames(ChiSqDiff) <- c(``Deviance'',``df'',``p-value'')
ChiSqDiff
```

```
> ChiSqDiff
      Deviance df p-value
[1,]   0.0315  1  0.8591
```

The degrees of freedom for complete symmetry model is 6, whereas for conditional symmetry model is 5; the degrees of freedom for the Chi-Square difference test is the difference of the two model with df = 6 − 5 = 1. The deviance of complete symmetry model is 2.7739 and for conditional symmetry is 2.7424, the difference in deviances is 0.0315, resulted in p-value of 0.8591. Since p-value is more than 0.05, we do not reject $H_0 : = \gamma = 1$. Hence, complete symmetry model is preferred. Comparing the fit statistics AIC and BIC also show that complete symmetry is preferred. The estimated value of γ for conditional symmetry model is equal to 0.9844 exp(−0.0158). The value is very close to one. This result indicates that conditional symmetry does not make much practical sense as it is very close to complete symmetry model for the value of γ is almost equal to one.

10.4.2   Chi-Square Difference Test—Complete Point Symmetry and Inclined Point Symmetry Model

Complete point symmetry model states that $H_p : \pi_{ij} = \pi_{i^*j^*}$, and inclined point symmetry model states that $H_{lp} : \pi_{ij} = r\pi_{i^*j^*}$. When $r = 1$, inclined point symmetry model reduces to complete point symmetry model.

The R syntax to carry out Chi-Square difference test of $H_0 : = r = 1$ for the above table is as follows.

```
PData <- c(42, 37, 55, 99,
           42, 32, 32, 87,
           60, 25, 33, 57,
           73, 42, 27, 43)
```

```
P  <- c(1, 3, 4, 5,
        8, 2, 6, 7,
        7, 6, 2, 8,
        5, 4, 3, 1)
P <- as.factor(P)
Psi <- c( 3, 1, 1, 1,
          2, 3, 1, 1,
          2, 2, 3, 1,
          2, 2, 2, 3)
CompletePoint <- glm(PData ~ P, family=poisson)
InclinedPoint  <- glm(PData ~ P + Psi, family=poisson)
df         <- CompletePoint$df.residual-InclinedPoint$df.residual
deviance <- CompletePoint$deviance-InclinedPoint$deviance
p          <- pchisq(deviance, df, lower.tail=F)
ChiSqDiff <- cbind(round(deviance,4), df, round(p,4))
colnames(ChiSqDiff) <- c("Deviance","df","p-value")
ChiSqDiff
```

```
> ChiSqDiff
     Deviance df p-value
[1,]  15.161  1   1e-04
```

The difference in degrees of freedom for complete point symmetry model and inclined point symmetry is one. The difference in G^2 is 15.161. Since p-value is less than 0.05, we reject $H_0 : r = 1$ and conclude that inclined point symmetry model is preferred. The fit statistics AIC and BIC also show that inclined point symmetry is a better fitted model. The estimated value of r is equal to 0.7330, $\exp(-0.3107)$ which is not closed to one.

Exercise

10.1. Table 10.17 tabulates literacy skills and leadership skills of 353 employed persons. Fit complete point, inclined point, and proportional point symmetry models and carry out Chi-Square difference test to examine the best-fitted model.

10.2. Table 10.18 tabulates father's and son's economic status of 3793 paired of fathers and sons. Fit independence, fixed distance, and variable distance models and carry out Chi-Square difference test to examine the best-fitted model.

10.3. Table 10.19 tabulates the rating of two raters 1 and 2 on a group of examinees sat for a test. Both raters rate them into four categories: excellent, good, average, and below average. Fit complete symmetry, conditional symmetry, and diagonal parameters models and carry out Chi-Square difference test to examine the best-fitted model.

10.4. Table 10.20 tabulates father and son occupation status. Fit conditional symmetry, linear diagonal parameters models, and 2-ratios parameters symmetry and carry out Chi-Square difference test to examine the best-fitted model.

Table 10.17 Literacy skills × leadership skills

Literacy skills	Leadership skills			
	1	2	3	4
1	99	9	5	11
2	22	18	6	17
3	33	11	43	11
4	22	11	18	17

Table 10.18 Occupational status of father and son

Father's status	Son's status				
	(1)	(2)	(3)	(4)	(5)
(1)	35	42	20	30	16
(2)	34	180	83	130	67
(3)	17	89	234	192	100
(4)	23	128	172	785	401
(5)	9	48	64	294	600

Table 10.19 Rating of two raters

Rater 2	Rater 1				
	Excellent	Good	Average	Below average	Bad
Excellent	30	20	32	11	53
Good	31	40	31	12	9
Average	15	64	32	9	22
Below average	35	7	16	78	15
Bad	25	20	15	32	60

Table 10.20 Father and son occupation status

Son occupation status	Father occupation status					
	1	2	3	4	5	6
1	33	9	6	2	2	1
2	18	25	11	9	4	2
3	16	21	68	20	12	8
4	6	23	40	74	35	14
5	10	14	30	71	132	55
6	2	9	24	35	110	120

10.5. Table 10.21 tabulates the scores of rater A and rater B of 923 assessments. Fit odds symmetry I, odds symmetry II, and quasi odds symmetry model and carry out Chi-Square difference test to examine the best-fitted model.
10.6. Table 10.22 is from Bishop et al. (1975). Fit the data with the models discussed from Chaps. 4, 5, 6, and 7.
10.7. Table 10.23 is from Bishop et al. (1975). Fit the data with the models discussed from Chaps. 4, 5, 6, and 7.
10.8. Table 10.24 is from Agresti (1984). Fit the data with the models discussed from Chaps. 4, 5, 6, and 7.
10.9. Two raters, Lim and Sheng rate 160 assignments and classify them into Grade A, B, and C. Fit complete symmetry, independence, quasi symmetry, and quasi independence model and comment on the results. Do Lim and Sheng agree with each other? (Table 10.25).

Table 10.21 Rater A × rater B

Rater B	Rater A					
	1	2	3	4	5	6
1	56	10	12	20	23	11
2	21	20	44	22	27	40
3	40	21	70	20	13	16
4	15	12	21	40	22	13
5	40	13	12	34	55	6
6	21	22	25	20	10	56

Table 10.22 Occupation status—British father–son mobility table

Father's status	Son's status				
	(1)	(2)	(3)	(4)	(5)
(1)	50	45	8	18	8
(2)	28	174	84	154	55
(3)	11	78	110	223	96
(4)	14	150	185	714	447
(5)	3	42	72	320	411

Source Lawal (2003)

Table 10.23 Occupation status—Danish father–son mobility table

Father's status	Son's status				
	(1)	(2)	(3)	(4)	(5)
(1)	18	17	16	4	2
(2)	24	105	109	59	21
(3)	23	84	289	217	95
(4)	8	49	175	348	198
(5)	6	8	69	201	246

Source Lawal (2003)

Table 10.24 Occupation
status—Danish father–son
mobility table

Father's status	Son's status				
	(1)	(2)	(3)	(4)	(5)
(1)	50	45	8	18	8
(2)	28	174	84	154	55
(3)	11	78	110	223	96
(4)	14	150	185	714	447
(5)	3	42	72	320	411

See Agresti (1984)

Table 10.25 Rating of sheng
and lim

Lim	Sheng		
	A	B	C
A	24	8	13
B	8	13	11
C	10	9	64

References

Agresti (1984). Analysis of ordinal categorical data. New York: John Wiley & Sons.

Bishop, Y. M., Fienberg, S. E., & Holland, P. W. (1975). Discrete Multivariate Analysis: Theory and Applications. Springer.

Lawal, H. B. (2000). Implementing point-symmetry models for square contingency tables having ordered categories in SAS. *Journal of the Italian Statistical Society, 9*, 1–22.

Lawal, H. B. (2003). Categorical Data Analysis with SAS and SPSS Applications. Lawrence Erlbaum Associates, Publishers.

Lawal, H. B. (2004). Using a GLM to decompose the symmetry model in square contingency tables with ordered categories. *Journal of Applied Statistics, 31*(3), 279–303.

Simonoff, J. S. (2003). Analyzing Categorical Data. Springer.

Index

© Springer Nature Singapore Pte Ltd. 2017
T.K. Tan, *Doubly Classified Model with R*,
https://doi.org/10.1007/978-981-10-6995-6

Printed in the United States
By Bookmasters